Econometrics and Data Science

Apply Data Science Techniques
to Model Complex Problems
and Implement Solutions
for Economic Problems

Tshepo Chris Nokeri

Apress®

Econometrics and Data Science

Tshepo Chris Nokeri
Pretoria, South Africa

ISBN-13 (pbk): 978-1-4842-7433-0   ISBN-13 (electronic): 978-1-4842-7434-7
https://doi.org/10.1007/978-1-4842-7434-7

Managing Director, Apress Media LLC: Welmoed Spahr
Acquisitions Editor: Celestin Suresh John
Development Editor: Laura Berendson
Coordinating Editor: Aditee Mirashi

Cover designed by eStudioCalamar

Cover image designed by Freepik (www.freepik.com)

Distributed to the book trade worldwide by Springer Science+Business Media New York, 1 New York Plaza, Suite 4600, New York, NY 10004-1562, USA. Phone 1-800-SPRINGER, fax (201) 348-4505, e-mail orders-ny@ springer-sbm.com, or visit www.springeronline.com. Apress Media, LLC is a California LLC and the sole member (owner) is Springer Science + Business Media Finance Inc (SSBM Finance Inc). SSBM Finance Inc is a **Delaware** corporation.

For information on translations, please e-mail booktranslations@springernature.com; for reprint, paperback, or audio rights, please e-mail bookpermissions@springernature.com.

Apress titles may be purchased in bulk for academic, corporate, or promotional use. eBook versions and licenses are also available for most titles. For more information, reference our Print and eBook Bulk Sales web page at http://www.apress.com/bulk-sales.

Any source code or other supplementary material referenced by the author in this book is available to readers on GitHub via the book's product page, located at www.apress.com/978-1-4842-7433-0. For more detailed information, please visit http://www.apress.com/source-code.

Printed on acid-free paper

I dedicate this book to my family and everyone who merrily played influential roles in my life, i.e., Professor Chris William Callaghan and Mrs. Renette Krommenhoek from the University of the Witwatersrand, among others I did not mention.

Table of Contents

About the Author

 **Tshepo Chris Nokeri** harnesses advanced analytics and artificial intelligence to foster innovation and optimize business performance. In his functional work, he delivered complex solutions to companies in the mining, petroleum, and manufacturing industries. He earned a Bachelor's degree in Information Management and then graduated with an Honour's degree in Business Science at the University of the Witwatersrand on a TATA Prestigious Scholarship and a Wits Postgraduate Merit Award. He was also unanimously awarded the Oxford University Press Prize. He is the author of *Data Science Revealed* and *Implementing Machine Learning in Finance,* both published by Apress.

About the Technical Reviewer

Pratibha Saha is an economics graduate currently working as an Economist Analyst-Consultant at Arthashastra Intelligence. She is trained in econometrics, statistics, and finance with interests in machine learning, deep learning, AI, et al.

She is motivated by the idea of problem solving with a purpose and strongly believes in diversity facilitating tech to supplement socially aware decision making. She finds technology to be a great enabler and understands the poignancy of data-driven solutions. By investigating the linkages of tech, AI, and social impact, she hopes to use her skills to propel these solutions.

Additionally, she is a feline enthusiast and loves dabbling in origami.

Find her on LinkedIn at `https://www.linkedin.com/in/pratibha-saha-8089a3192/` and on GitHub at `https://github.com/Pratsa09`

Acknowledgments

Writing a single-authored book is demanding, but I received firm support and active encouragement from my family and dear friends. Many heartfelt thanks to the Apress Publishing team for all their support throughout the writing and editing processes. Last, my humble thanks to all of you for reading this; I earnestly hope you find it helpful.

Introduction

This book bridges the gap between econometrics and data science techniques. It introduces a holistic approach to satisfactorily solving economic problems from a machine learning perspective. It begins by discussing the practical benefits of harnessing data science techniques in econometrics. It then clarifies the key concepts of variance, covariance, and correlation, and covers the most common linear regression model, called ordinary least-squares. It explains the techniques for testing assumptions through residual analysis, including other evaluation metrics (i.e., mean absolute error, mean squared error, root mean squared error, and R2). It also exhibits ways to correctly interpret your findings. Following that, it presents an approach to tackling series time data by implementing an alternative model to the dominant time series analysis models (i.e., ARIMA and SARIMA), called the additive model. That model typically adds non-linearity and smooth parameters.

The book also introduces ways to capture non-linearity in economic data by implementing the most prevalent binary classifier, called logistic regression, alongside metrics for evaluating the model (i.e., confusion matrix, classification report, ROC curve, and precision-recall curve). In addition, you'll learn about a technique for identifying hidden states in economic data by implementing the Hidden Markov modeling technique, together with an approach for realizing mean and variance in each state. You'll also learn how to categorize countries grounded on similarities by implementing the most common cluster analysis model, called the K-Means model, which implements the Euclidean distance metric.

The book also covers the practical application of deep learning in econometrics by showing key artificial neural networks (i.e., restricted Boltzmann machine, multilayer perceptron, and deep belief networks), including ways of adding more complexity to networks by expanding hidden layers. Then, it familiarizes you with a method of replicating economic activities across multiple trials by implementing the Monte Carlo simulation technique. The books concludes by presenting a standard procedure for testing causal relationships among variables, including the mediating effects of other variables in those relationships, by implementing structural equation modeling (SEM).

This book uses Anaconda (an open-source distribution of Python programming) to prepare examples. Before exploring the contents of this book, you should understand the basics of economics, statistics, Python programming, probability theories, and predictive analytics. The following list highlights some Python libraries that this book covers.

- Wdata for extracting data from the World Bank database
- Scikit-Learn for building and validating key machine learning algorithms
- Keras for high-level frameworks for deep learning
- Semopy for performing structural equation modeling
- Pandas for data structures and tools
- NumPy for arrays and matrices
- Matplotlib and Seaborn for recognized plots and graphs

This book targets beginners to intermediate economists, data scientists, and machine learning engineers who want to learn how to approach econometrics problems from a machine learning perspective using an array of Python libraries.

CHAPTER 1

Introduction to Econometrics

This chapter explains data science techniques applied to the field of *econometrics*. To begin, it covers the relationship between economics and quantitative methods, which paves the way for the econometrics field. It also covers the relevance of econometrics in devising and revising the economic policies of a nation. It then summarizes machine learning, deep learning, and structural equation modeling. To conclude, it reveals ways to extract macroeconomic data using a standard Python library.

Econometrics

Econometrics is a social science subclass that investigates broad business activities at the macro level, i.e., at the country, region, or continent level. It is an established social science field that employs statistical models to investigate theoretical claims about macroeconomic phenomena. Figure 1-1 is a simplification of econometrics. Organizations like the statistical bureau capture economic activities across time, which they make available to the public. Practitioners, such as economists, research analysts, and statisticians alike, extract the data and model it using algorithms grounded on theoretical frameworks in order to make future predictions.

© Tshepo Chris Nokeri 2022
T. C. Nokeri, *Econometrics and Data Science*, https://doi.org/10.1007/978-1-4842-7434-7_1

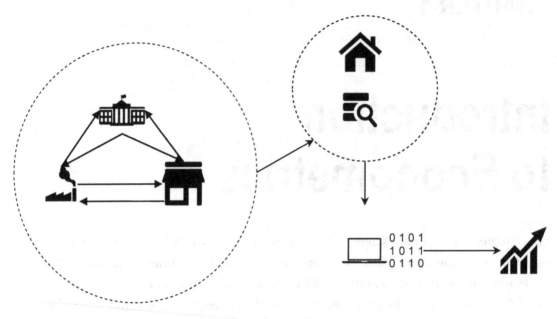

Figure 1-1. *Econometrics*

Before you proceed with the contents of this book, be sure that you understand the basic concepts that relate to economics and statistics.

Economic Design

Economic design is grounded on the notion that if we can accurately estimate macroeconomic phenomenon, we can devise mechanisms that help manage it. As mentioned, there are several well-established organizations from which one can extract factual macroeconomic data. Note that we cannot estimate the whole population, but we can use a sample (a representative of the population) because there are errors in statistical estimations. Because there is a pool of reliable macroeconomic data sources, we can apply the data and investigate consistent patterns by applying quantitative models to make sense of an economy. When we are confident that a model estimates what we intend it to estimate and does so exceptionally, we can apply such a model to predict economic events. Remember that the primary purpose of a scientific enterprise is to predict events and control underlying mechanisms by applying quantitative models.

Econometrics uses statistical principles to estimate the parameters of a population, but the ultimate litmus test is always economic ideology. Only economic theory can validate/invalidate the results, which can be further used to determine causation/

correlation, etc. It should be apparent that politics occupies a paramount role in modern life. At most, the political sentiments typically accompany a firm belief about the economy and how it ought to be. Such beliefs might not reflect economic reality. When the considered belief about the economy is absurd, there is no way of combating pressing societal problems with devised solutions. To satisfactorily solve an economic problem, you must have a logical view; otherwise, feelings, standard assumptions, and authoritarian knowledge dilute your analysis of an economy.

In summary, policymakers apply econometrics to devise and revise economic policies so that they can correctly solve economic problems. This entails that they investigate historical economic events, develop complex quantitative models, and apply findings of those models (provided they are reliable) to drive economic policies. Econometrics is an approach for finding answers to questions that relate to the economy. Policymakers who are evidence-oriented drive policymaking initiatives by applying factual data rather than depending on political and economic ideologies.

Understanding Statistics

Statistics is the field concerned with discovering consistent patterns in raw data to derive a logical conclusion regarding a recognized phenomenon. It involves investigating the central tendency (the mean value) and dispersion of data (the standard deviation) and then studying theoretical claims about the phenomenon by applying quantitative models. In addition, business institutions apply it in ad hoc reporting, research, and business process controls. Researchers, in addition, apply statistics in fields like natural sciences, physical sciences, chemistry, engineering, and social sciences, among other fields. It is the backbone of quantitative research.

Machine Learning Modeling

There is a link between statistics and machine learning. In this book, we consider machine learning an extension of statistics that incorporates techniques from fields like computer science. Machine learning methods derive from statistical principles and methods. We approach machine learning problems with "applications" and "automation" in mind. With machine learning, the end goal is not to derive some conclusion but to automate monotonous tasks and determine replicable patterns for those autonomous tasks. Figure 1-2 shows how quantitative models operate.

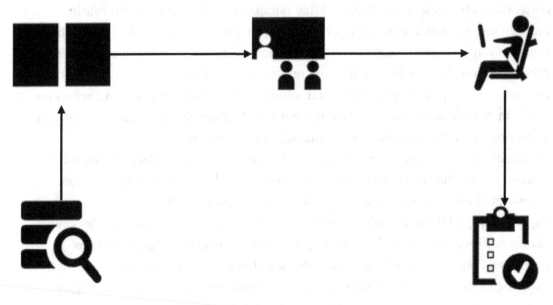

Figure 1-2. *Fundamental machine learning model*

Figure 1-2 demonstrates the basic machine learning model flow. Initially, we extract the data from a database, then preprocess and split it. This is followed by modeling the data by applying a function that receives a predictor variable and operates it to generate an output value. A variable represents a process that we can observe and estimate. It is common practice in machine learning to deploy models as web applications or as part of web applications.

Deep Learning Modeling

Deep learning applies artificial neural networks (a reciprocal web of nodes) that imitate the human neural structure. Artificial neural networks are a group of nodes that receive input values in the input layer, transform them to the subsequent hidden layer (a layer between the input and output layer), which transforms them and allots varying *weights* (vector parameters that determine the extent of influence input values have on output values) and *bias* (a balance value which is 1). It is a subclass of machine learning that combats some difficulties that we encounter with conventional quantitative models. For instance, the vanishing gradient problem—a case in which the gradient is minimal at the initial phase of the training process and increases as we include more data. There are other types of artificial neural networks, i.e. Restricted Boltzmann Machine—a shallow network between the hidden layer and output layer, Multilayer Perceptron—a neural network with over two hidden layers, Recurrent Neural Network—a neural network for solving sequential data, and Convolutional Neural Network—a neural network for dimension reduction, frequently applied in computer vision. This covers the Restricted Boltzmann Machine and Multilayer Perceptron. Figure 1-3 shows a Multilayer Perceptron classifier.

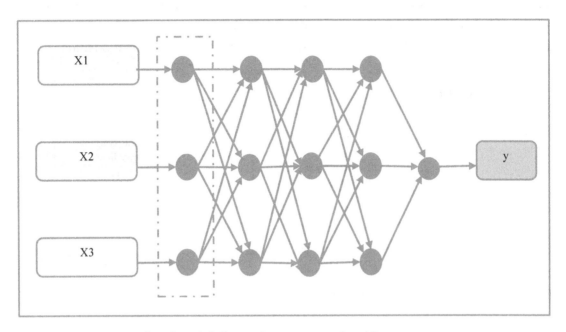

Figure 1-3. *Example of a Multilayer Perceptron classifier*

Figure 1-3 shows that the Multilayer Perceptron classifier is composed of an input layer that retrieves input values (X1, X2, and X3) and conveys them to the first hidden layer. That layer then retrieves the values and transforms them by applying a function (in our case, the Sigmoid function). It conveys an output value, which is then conveyed to the second hidden layer, which also retrieves the input values. The process reiterates—it transforms values and conveys them to the output layer and produces an output value represented as Y in Figure 1-3. We recognized the training process that networks apply to learn the structure of the data, recognized as backward propagation (updating weights in reverse). Chapter 8 covers deep learning.

Structural Equation Modeling

The structural equation model includes a set of models that determine the nature of causal relationships among sets of variables. It includes factor analysis, path analysis, and regression analysis. It helps us investigate mediating relationships, so we can detect how the presence of other variables weakens or strengthens the nature of the structural relationship between the predictor variable and the response variable. Figure 1-4 shows a hypothetical framework that outlines direct and indirect structural relationships.

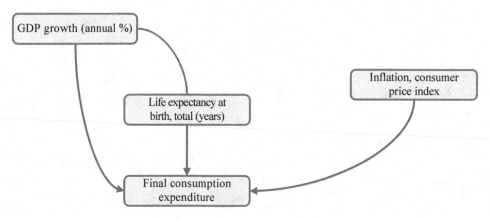

Figure 1-4. *Fundamental structural equation model*

Figure 1-4 demonstrates a hypothetical framework representing the structural relationship between GDP per capita growth (as an annual percentage), inflation, consumer price index (as a percentage), and final consumption expenditure (in current U.S. dollars). In addition, it highlights the mediating effects of life expectancy on the relationship between GDP per capita growth and final consumption expenditure. Chapter 10 covers structural equation modeling.

Macroeconomic Data Sources

There are several libraries that are used to extract macroeconomic data. This book uses one of the more prominent libraries, called wbdata. This library extracts data from the World Bank database[1]. Alternatively, you can extract the data from the World Bank

---

[1] Indicators | Data (worldbank.org)

website directly. In addition, there are other macroeconomic sources you can use, such as the St. Louis Fed (Federal Reserve Economic) database[2] and the International Monetary Fund database[3], among others.

This book uses the `world-bank-data` library as the principal library because it offers a wide range of social indicators. Before you proceed, ensure that you install the `world-bank-data` library. This will make the process of developing quantitative models much simpler, as you will not have to write considerable chunks of code. To install the library in the Python environment, use `pip install wbdata`. Provided you are using the Anaconda environment, use `conda install wbdata`. At the point of printing this book, the version of the library was v0.3.0. Listing 1-1 shows how to retrieve the macroeconomic data.

Listing 1-1. Loading Data from the World Bank Library

```
import wbdata
country = ["USA"]
indicator = {"FI.RES.TOTL.CD":"gdp_growth"}
df = wbdata.get_dataframe(indicator, country=country, convert_date=True)
```

`wbdata` extracts the data and loads it into a `pandas` dataframe. Figure 1-5 demonstrates the `wbdata` workflow.

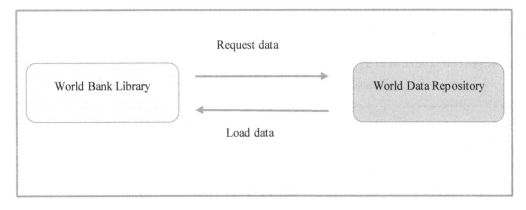

Figure 1-5. *World Bank library workflow*

---

[2] Federal Reserve Economic Data | FRED | St. Louis Fed (stlouisfed.org)
[3] IMF Data

Extracting data from the wbdata library requires that you specify the country ID. Given that the World Bank includes several countries, it is burdensome to know the IDs of all of them. The most convenient way to find a country's ID is to search for it by name (see Listing 1-2). For this example, we entered China, and it returned Chinese countries, including their IDs.

Listing 1-2. Searching for a Country ID

```
wbdata.search_countries("China")
id     name
----   --------------------
CHN    China
HKG    Hong Kong SAR, China
MAC    Macao SAR, China
TWN    Taiwan, China
```

Extracting data from the wbdata library requires that you specify the economic indicator's ID as well. Given that the World Bank includes several macroeconomic indicators, it is burdensome to know the IDs of all the indicators. The most convenient way to find an indicator's ID is to search for it by name (see Listing 1-3). For this example, we entered inflation and it returned all indicators that contain the word "inflation," including their IDs.

Listing 1-3. Searching for Macroeconomic Data

```
wbdata.search_indicators("inflation")
id                     name
--------------------   ----------------------------------------------------
FP.CPI.TOTL.ZG         Inflation, consumer prices (annual %)
FP.FPI.TOTL.ZG         Inflation, food prices (annual %)
FP.WPI.TOTL.ZG         Inflation, wholesale prices (annual %)
NY.GDP.DEFL.87.ZG      Inflation, GDP deflator (annual %)
NY.GDP.DEFL.KD.ZG      Inflation, GDP deflator (annual %)
NY.GDP.DEFL.KD.ZG.AD   Inflation, GDP deflator: linked series (annual %)
```

The wbdata library includes several data sources, like World Development Indicators, Worldwide Governance Indicators, Subnational Malnutrition Database, International Debt Statistics, and International Debt Statistics: DSSI, among others. This book focuses predominantly on sources that provide economic data. It also covers social indicators. Listing 1-4 demonstrates how to retrieve indicator sources using `wbdata.get_source()` (see Table 1-1).

Listing 1-4. Retrieving the World Bank Sources

```
sources = wbdata.get_source()
sources
```

Table 1-1. *World Bank Sources*

	ID	Last Updated	Name	Code	Description	URL	Data Availability	Metadata Availability	Concepts
0	1	2019-10-23	Doing Business	DBS			Y	Y	3
1	2	2021-05-25	World Development Indicators	WDI			Y	Y	3
2	3	2020-09-28	Worldwide Governance Indicators	WGI			Y	Y	3
3	5	2016-03-21	Subnational Malnutrition Database	SNM			Y	Y	3
4	6	2021-01-21	International Debt Statistics	IDS			Y	Y	4
...	...	...	...	...	...	...	...	...	...
60	80	2020-07-25	Gender Disaggregated Labor Database (GDLD)	GDL			Y	N	4
61	81	2021-01-21	International Debt Statistics: DSSI	DSI			Y	N	4

(continued)

Table 1-1. (*continued*)

	ID	Last Updated	Name	Code	Description	URL	Data Availability	Metadata Availability	Concepts
62	82	2021-03-24	Global Public Procurement	GPP			Y	N	3
63	83	2021-04-01	Statistical Performance Indicators (SPI)	SPI			Y	Y	3
64	84	2021-05-11	Education Policy	EDP			Y	Y	3

Table 1-1 outlines the source ID, name, code, availability, metadata availability, concepts, and the last date of update. Listing 1-5 shows how to retrieve the topics (see Table 1-2). Each topic has its own ID.

Listing 1-5. Retrieve Topic

```
wbdata.get_topic()
```

Table 1-2. *World Bank Topic*

	ID	Value	Source Note
0	1	Agriculture & Rural Development	For the 70 percent of the world's poor who liv...
1	2	Aid Effectiveness	Aid effectiveness is the impact that aid has i...
2	3	Economy & Growth	Economic growth is central to economic develop...
3	4	Education	Education is one of the most powerful instrume...
4	5	Energy & Mining	The world economy needs ever-increasing amount...
5	6	Environment	Natural and man-made environmental resources –...
6	7	Financial Sector	An economy's financial markets are critical to...
7	8	Health	Improving health is central to the Millennium ...
8	9	Infrastructure	Infrastructure helps determine the success of ...
9	10	Social Protection & Labor	The supply of labor available in an economy in...

(continued)

Table 1-2. (*continued*)

	ID	Value	Source Note
10	11	Poverty	For countries with an active poverty monitorin...
11	12	Private Sector	Private markets drive economic growth, tapping...
12	13	Public Sector	Effective governments improve people's standar...
13	14	Science & Technology	Technological innovation, often fueled by gove...
14	15	Social Development	Data here cover child labor, gender issues, re...
15	16	Urban Development	Cities can be tremendously efficient. It is ea...
16	17	Gender	Gender equality is a core development objectiv...
17	18	Millennium development goals	
18	19	Climate Change	Climate change is expected to hit developing c...
19	20	External Debt	Debt statistics provide a detailed picture of ...
20	21	Trade	Trade is a key means to fight poverty and achi...

Table 1-2 outlines source ID, values of the topic, and source notes. The wbdata library encompasses a broad range of topics from fields like health, economics, urban development, and other social science-related fields.

Context of the Book

Each chapter of this book starts by covering the underlying concepts of a particular model. The chapters show ways to extract macroeconomic data for exploration, including techniques to ensure that the structure of the data is suitable for a chosen model and meets preliminary requirements. In addition, the chapters reveal possible ways of establishing a hypothetical framework and testable hypotheses. They discuss how to investigate hypotheses by employing a quantitative model that operates a set of variables to generate output values.

For each model in this book, there are ways to evaluate it. Each chapter also includes visuals that will help you better understand the structure of the data and the results.

Practical Implications

This book expands on the present body of knowledge on econometrics. It covers ways through which you can apply data science techniques to discover patterns in macroeconomic data and draw meaningful insights. It intends on accelerating evidence-based economic design—devising and revising economic policies based on evidence that we derive from quantitative-driven models. This book is for professionals who seek to approach some of the world's most pressing problems by applying data science and machine learning techniques. In summary, it will enable you to detect *why* specific social and economic activities occur, and help you predict the likelihood of future activities occurring. The book assumes that you have some basic understanding of key concepts of statistics and economics.

CHAPTER 2

Univariate Consumption Study Applying Regression

This chapter introduces the standard univariate (or simple) linear regression model, called the ordinary least-squares model, which estimates the intercept and slope while diminishing the residuals (see Equation 2-1). It applies the model to determine the relationship between the interest rates that U.S. banks charge for lending and the market value of goods and services that U.S. households consume annually. It includes ways of conducting covariance analysis, correlation analysis, model development, cross-validation, hyperparameter optimization, and model performance analysis.

The ordinary least-squares model is one of the most common parametric methods. It establishes powerful claims regarding the data—it expects *normality* (values of a variable saturating the mean value) and *linearity* (an association between an independent variable and a dependent variable). This chapter uses the most common parametric method, called the *ordinary least-squares model,* to investigate the association between the predictor variable (the independent variable) and the response variable (the dependent variable). It's based on a straight-line equation (see Figure 2-1).

© Tshepo Chris Nokeri 2022
T. C. Nokeri, *Econometrics and Data Science,* https://doi.org/10.1007/978-1-4842-7434-7_2

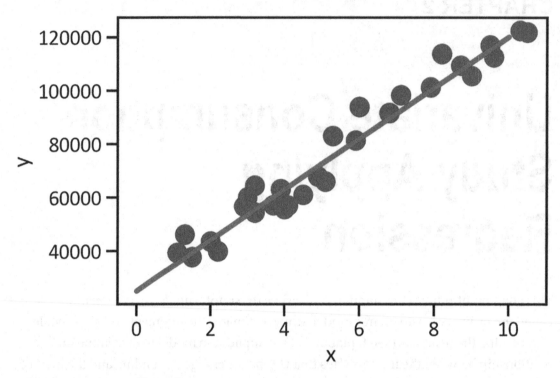

Figure 2-1. *Line of best fit*

Figure 2-1 shows a straight line in red and the independent data points in green—the line cuts through the data points. Equation 2-1 shows the ordinary least-squares equation.

$$\hat{y} = \hat{\beta}_0 + \hat{\beta}_1 \hat{X}_1 + \hat{\varepsilon}_i \qquad \text{(Equation 2-1)}$$

Where $\hat{y}$ is the predicted response variable (the expected the U.S. final consumption expenditure in this example), $\hat{\beta}_0$ represents the intercept—representing the mean value of the response variable (the U.S. final consumption expenditure in U.S. dollars for this example), $\hat{X}_1$ represents the predictor variable (the U.S. lending interest rate in this example), and 1 $\hat{\beta}_1$ is the slope—representing the direction of the relationship between X (the U.S. lending interest rate) and the final consumption expenditure (in current U.S. dollars). Look at the straight red line in Figure 2-1—the slope is positive). Finally, $\hat{\varepsilon}_i$ represents the error in terms (refer to Equation 2-2).

$$\hat{\varepsilon}_i = y_i - \hat{y}_i \qquad \text{(Equation 2-2)}$$

Where ε_i is the error in term (also called the *residual term*)—representing the difference between yi (the actual U.S. final consumption expenditure) and $\hat{y}\ i$ (the predicted U.S. final consumption expenditure).

There is a difference between variables with a hat/caret (which are sample regression functions) and without one (population regression functions). We estimate those containing a hat/caret from a sample of the population, rather than from the entire population.

Context of This Chapter

This chapter uses the ordinary least-squares regression model to determine the linear relationship between the predictor variable (the U.S. lending interest rate as a percentage) and the response variable (final consumption expenditure in current U.S. dollars). Table 2-1 outlines the macroeconomic indicators for this chapter.

Table 2-1. *The U.S. Macroeconomic Indicators for This Chapter*

Code	Title
FR.INR.LEND	Lending interest rate (as a percentage)
NY.GDP.MKTP.KD.ZG	Final consumption expenditure (in current U.S. dollars)

Theoretical Framework

Figure 2-2 shows the relationship that this chapter explores. It establishes the research hypothesis.

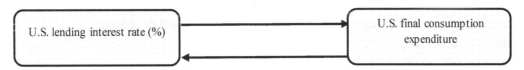

Figure 2-2. *Theoretical framework*

HYPOTHESES

H_0: There is no significant difference between the U.S. lending interest rate (%) and the final consumption expenditure (in current U.S. dollars).

H_A: There is a significant difference between U.S. social contributions and the final consumption expenditure (in current U.S. dollars).

The research hypothesis seeks to determine whether a change in the U.S. lending interest rate influences the final consumption expenditure (in current U.S. dollars).

Lending Interest Rate

The lending interest rate estimates the rate that private banks charge as interest for short-term and mid-term loans. We express the estimate as an annual percentage. Figure 2-3 demonstrates the rate that U.S.-based private banks charged as interest for short-term and mid-term financing from 1960 to 2020. Before you proceed, be sure that you have the Matplotlib library installed in your environment. To install the Matplotlib library in a Python environment, use `pip install matplotlib`. Equally, to install the library in a Conda environment, use `conda install -c conda-forge matplotlib`. See Listing 2-1.

Listing 2-1. The U.S. Lending Interest Rate

```
import wbdata
from matplotlib.pyplot import
%matplotlib inline
country  = ["USA"]
indicator = {"FP.CPI.TOTL.ZG":"lending_rate"}
inflation_cpi = wbdata.get_dataframe(indicator, country=country,
   convert_date=True)
inflation_cpi.plot(kind="line",color="green",lw=4)
plt.title("The U.S. lending interest rate (%)")
plt.ylabel("Lending interest rate (%)")
plt.xlabel("Date")
plt.legend(loc="best")
plt.show()
```

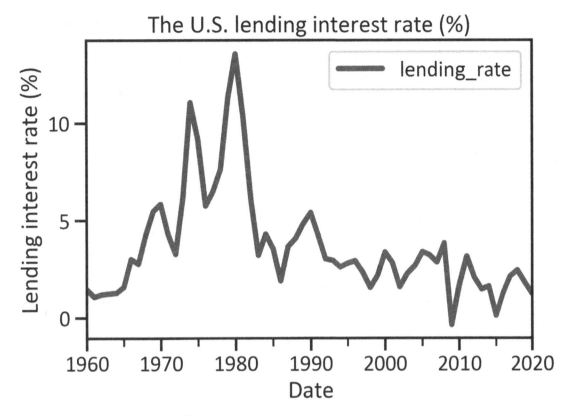

Figure 2-3. *The U.S. lending interest rate*

Figure 2-3 demonstrates that from 1960 to 1980, the rate that U.S.-based banks charged as interest for private short-term and mid-term financing grew from 1.45% to 13.55% (the highest peak). Following that, in the early 1980s, the rate sharply declined, then it remained stagnant. It also shows that the lending interest rate reached its lowest point in 2008. In summary, acquiring debt from U.S.-based banks was more expensive in the late 1970s, and relatively cheap in 2008.

Final Consumption Expenditure (in Current U.S. Dollars)

The final consumption expenditure is the market value of general goods and services that households in an economy purchase (see Listing 2-2). Figure 2-4 demonstrates the U.S. final consumption expenditure from 1960 to 2020.

Listing 2-2. The U.S. Final Consumption Expenditure (Current U.S. Dollars)

```
country = ["USA"]
indicator = {"NE.CON.TOTL.CD":"final_consumption"}
final_consumption = wbdata.get_dataframe(indicator, country=country,
convert_date=True)
final_consumption.plot(kind="line",color="orange",lw=4)
plt.title("The U.S. FCE")
plt.ylabel("FCE")
plt.xlabel("Date")
plt.legend(loc="best")
plt.show()
```

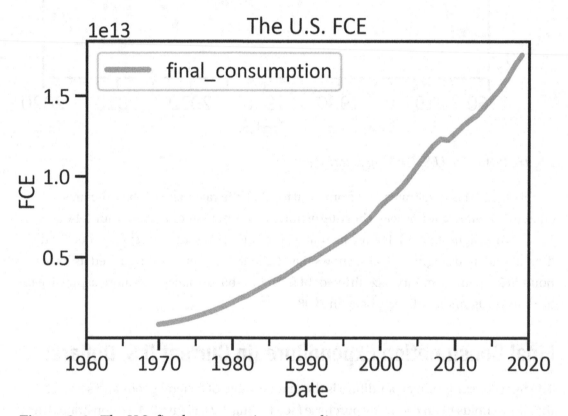

Figure 2-4. *The U.S. final consumption expenditure*

Figure 2-4 shows that there has been an uninterruptible upswing in the market value of general goods and services that U.S. households purchased since 1970.

The Normality Assumption

A normal distribution is also called a Gaussian distribution. This book uses these terms interchangeably. This distribution shows saturation data points around the mean data point. Ordinary least-squares regression models assume data points saturate the mean data point, thus you must detect normality before training the model on the U.S. lending interest rate and the final consumption expenditure data.

Normality Detection

Normality detection involves investigating the central tendency of data points. If the U.S. lending interest rate and the final consumption expenditure data saturates the mean data point, you can summarize the data using the central data points. An additional implicit assumption is that the errors follow the normality assumption, which makes them easier to deal with. To detect normality, you must estimate the mean data point (see Equation 2-3).

$$\bar{x} = \frac{x_1 + x_2 x + x_3 \ldots x_i}{n} \qquad \text{(Equation 2-3)}$$

Where $\bar{x}$ is the mean value, x_1 is the first data point, x_2 is the second data point, and so forth, and n represents the total number of data points. You divide the count of data points by degrees of freedom. Alternatively, you can find the median data point (the central data point). To determine the dispersion, estimate the standard deviation (see Equation 2-4).

$$s = \sqrt{\frac{\sum_{i=1}^{n}(x_i - \bar{x})^2}{n-1}} \qquad \text{(Equation 2-4)}$$

After estimating the standard deviation, you square it to find the variance (see Equation 2-5).

$$s^2 = \frac{\sum_{i=1}^{n}(x_i - \bar{x})^2}{n-1} \qquad \text{(Equation 2-5)}$$

Listing 2-3 retrieves data relating to the U.S. lending interest and the final consumption expenditure data from the World Bank. See Table 2-2.

Listing 2-3. Load the U.S. Lending Interest Rate and Final Consumption Expenditure Data

```
country  = ["USA"]
indicators = {"FR.INR.LEND":"lending_rate",
              "NE.CON.TOTL.CD":"final_consumption"}
df = wbdata.get_dataframe(indicators, country=country, freq="M",
convert_date=False)
df.head()
```

Table 2-2. *The U.S. Lending Interest Rate and Final Consumption Expenditure Data*

Date	lending_rate	final_consumption
2020	3.544167	NaN
2019	5.282500	1.753966e+13
2018	4.904167	1.688457e+13
2017	4.096667	1.608306e+13
2016	3.511667	1.543082e+13

Table 2-2 shows that data points are missing from the data. Listing 2-4 substitutes the missing data points with the mean value.

Listing 2-4. Replacing Missing Data Points with the Mean Value

```
df["lending_rate"] = df["lending_rate"].fillna(df["lending_rate"].mean())
df["final_consumption"] = df["final_consumption"].fillna(df["final_
consumption"].mean())
```

Descriptive Statistics

There are several ways to visualize and summarize the distribution of data. The simplest way involves the use of a box plot. Box plots can also help detect normality. This plot confirms the location of the median data point. It also informs you about the length of the distribution tail, thus adequately supporting you in diagnosing outliers in the data. Figure 2-5 shows a box plot of the U.S. lending interest rate created by the code in Listing 2-5.

Listing 2-5. The U.S. Lending Interest Rate Distribution

```
df["lending_rate"].plot(kind="box",color="green")
plt.title("The U.S. lending interest rate (%)")
plt.ylabel("Values")
plt.show()
```

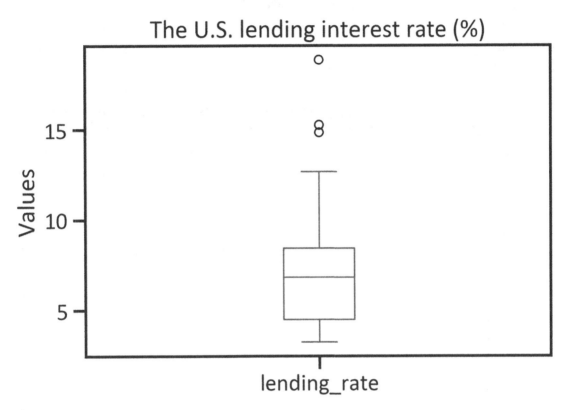

Figure 2-5. *The U.S. lending interest rate*

Figure 2-5 shows that there are three extreme values in the U.S. lending interest rate data. Listing 2-6 substitutes any outliers with the mean data point and determines the new distribution (see Figure 2-6). Before you proceed, be sure that you have the NumPy library installed in your environment. To install the NumPy library in a Python environment, use pip install numpy. Equally, to install the library in a Conda environment, use conda install -c anaconda numpy.

Listing 2-6. The U.S. Lending Interest Rate Distribution

```
import numpy as np
df['lending_rate'] = np.where((df["lending_rate"] > 14.5),df["lending_
rate"].mean(),df["lending_rate"])
df["lending_rate"].plot(kind="box",color="green")
plt.title("The U.S. lending interest rate (%)")
plt.ylabel("Values")
plt.show()
plt.show()
```

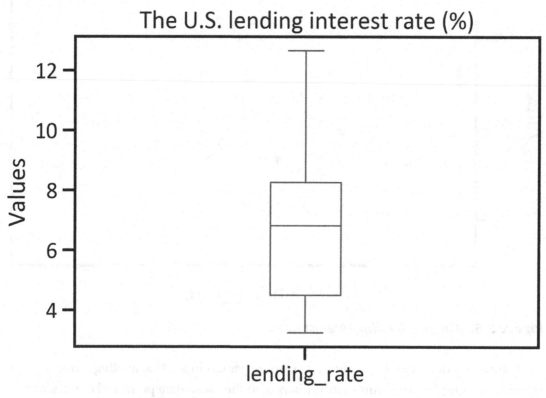

Figure 2-6. *The U.S. lending interest rate distribution*

Listing 2-7 returns Figure 2-7, which shows the distribution and outliers in the U.S. final consumption expenditure data.

Listing 2-7. The U.S. Final Consumption Expenditure Distribution

```
df["final_consumption"].plot(kind="box",color="orange")
plt.title("US FCE")
plt.ylabel("Values")
plt.show()
```

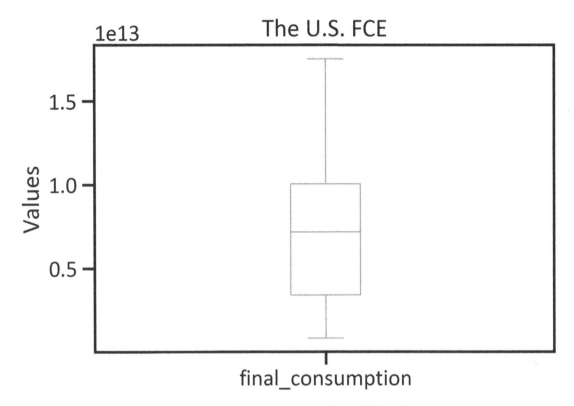

Figure 2-7. *The U.S. final consumption expenditure distribution*

Figure 2-7 shows there are no outliers in the U.S. final consumption expenditure data. The command in Listing 2-8 retrieves a comprehensive report relating to the central tendency and the dispersion of data points (see Table 2-3).

Listing 2-8. Descriptive Summary

```
df.describe()
```

Table 2-3. *Descriptive Summary*

	lending_rate	final_consumption
Count	61.000000	6.100000e+01
Mean	6.639908	7.173358e+12
Std	2.436502	4.580756e+12
Min	3.250000	8.395100e+11
25%	4.500000	3.403470e+12
50%	6.824167	7.173358e+12
75%	8.270833	1.006535e+13
Max	12.665833	1.753966e+13

Table 2-3 shows that:

- The mean value of the U.S. lending interest rate data is 6.639908 and the final consumption expenditure mean is 7.173358e+12.

- The data points of the U.S. lending interest rate deviate from the mean by 2.436502 and the final consumption expenditure data points deviate from the mean by 4.580756e+12.

Covariance Analysis

Covariance analysis involves estimating the extent to which variables vary with respect to each other. Equation 2-6 shows the covariance formula.

$$Covariance = \frac{\sum_{i=1}^{n}(x_i - \bar{x})^2(y_i - \bar{y})^2}{n-1} \qquad \text{(Equation 2-6)}$$

Where x_i represents independent data points of the lending interest rate (%) and $\bar{x}_i$ represents the mean value of the predictor variable. y_i represents the independent data points of the U.S. final consumption expenditure, and $\bar{y}_i$ represents the mean value of the U.S. final consumption expenditure. Listing 2-9 estimates the joint variability between the U.S. lending interest rate and the final consumption expenditure (see Table 2-4).

Listing 2-9. Covariance Matrix

```
dfcov = df.cov()
dfcov
```

Table 2-4. *Covariance Matrix*

	lending_rate	final_consumption
lending_rate	5.936544e+00	-7.092189e+12
final_consumption	-7.092189e+12	2.098332e+25

Table 2-4 shows that the U.S. lending interest rate variance is 5.936544e+00 and that the final consumption expenditure varies by 2.098332e+25. It shows you that the joint variability between the lending interest rate and the U.S. final consumption expenditure is -7.092189e+12. The next section provides an overview of correlation methods and explains which correlation method is used for this problem.

Correlation Analysis

Unlike covariance analysis, which shows how variables vary with respect to each other, *correlation analysis* estimates the dependency among variables. There are three principal correlation methods—the Pearson correlation method, which can estimate dependency among continuous variables, the Kendall method, which can estimate dependency among categorical variables, and the Spearman method, which also can estimate an association among categorical variables. Macroeconomic data is often continuous, so this chapter uses the Pearson correlation method. Most of the chapters in this book use this method, except for Chapter 5, which uses the Kendall method. Equation 2-7 estimates the covariance, then divides the estimate by the dispersion in the predictor variable and the response variable to retrieve a Pearson correlation coefficient.

$$r_{xy} = \frac{\sum_{i=1}^{n}(x_i - \bar{x})^2 (y_i - \bar{y})^2}{\sqrt{\sum_{i=1}^{n}(x_i - \bar{x})^2 \sum_{i=1}^{n}(y_i - \bar{y})^2}}$$ (Equation 2-7)

Where r_{xy} is the Pearson correlation coefficient. You can estimate that coefficient by dividing the covariance between the U.S. lending interest rate and the U.S. final consumption expenditure by the square root of the sum of the deviations. Listing 2-10 retrieves the Pearson correlation matrix (see Table 2-5).

Listing 2-10. Pearson Correlation Matrix

```
dfcorr = df.corr(method="pearson")
dfcorr
```

Table 2-5. *Pearson Correlation Matrix*

	lending_rate	final_consumption
lending_rate	1.000000	-0.635443
final_consumption	-0.635443	1.000000

Table 2-6 interprets the Pearson correlation coefficients outlined in Table 2-5.

Table 2-6. *Interpretation of Pearson Correlation Coefficients*

Relationship	Pearson Correlation Coefficient	Findings
The U.S. lending interest rate (%) and the final consumption expenditure (in current U.S. dollars)	-0.635443	There is an extreme negative correlation between the U.S. lending interest rate (%) and the final consumption expenditure (in current U.S. dollars).

Figure 2-8 shows the statistical difference between the U.S. lending interest rate and the final consumption expenditure. Before you proceed, be sure that you have the seaborn library installed in your environment. To install the seaborn library in a Python environment, use pip install seaborn. Equally, to install the library in a Conda environment, use conda install -c anaconda seaborn. See Listing 2-11.

Listing 2-11. Pairwise Scatter Plot

```
import seaborn as sns
sns.jointplot(x = "lending_rate", y="final_consumption", data=df,
kind="reg",color="navy")
plt.show()
```

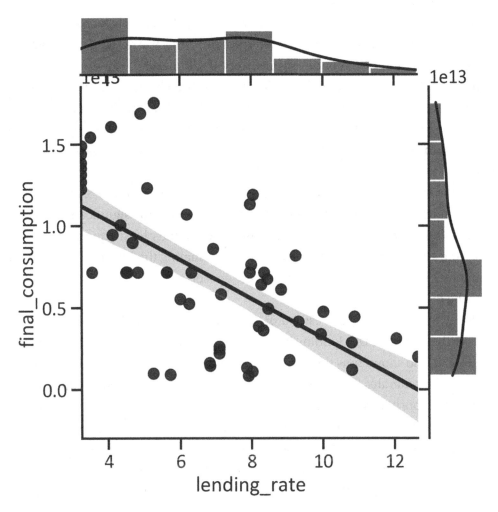

Figure 2-8. *The U.S. lending interest rate and the final U.S. consumption expenditure joint plot*

Figure 2-8 confirms that there is a negative correlation between the U.S. lending interest rate and the final consumption expenditure.

Ordinary Least-Squares Regression Model Development Using Statsmodels

The section covers the most commonly used regression model, called the ordinary least-squares model, which estimates the intercept and coefficients, while diminishing residuals (refer back to Equation 2-1).

Listing 2-12 converts the data to the required format. It begins by constructing an x array (the data points of the U.S. lending interest rate) and a y array (the data points of the U.S. final consumption expenditure in current U.S. dollars).

It then shapes the data so that the ordinary least-squares regression model better studies the data, then splits the data into training and test data, by applying the train_test_split() method. Lastly, it standardizes the data in such a way that the mean data point is 0 and the standard deviation is 1; it does this by applying the StandardScaler() method.

To test the claim that the U.S. lending interest rate influences the U.S. final consumption expenditure, you find the p-value to determine the significance of the relationship. In addition, you determine how the ordinary least-squares model expresses the amount of information it lost when estimating the future values of the final consumption expenditure. To further assess the chosen model's performance, you must estimate the R^2 score.

Table 2-7 outlines the type of estimator applied to test the significance of the relationship between the U.S. lending interest rate and the final U.S. consumption expenditure. It also shows how the model learns the available U.S. macroeconomic data, and how it generates future instances of the final U.S. consumption expenditure, including the errors it makes when estimating those instances. In addition, it details the extent to which the model explains how changes in the U.S. lending interest rate influence changes in the final U.S. consumption expenditure. In summary, it helps you decide whether you must accept the existence of the established macroeconomic phenomenon, including the degree to which you can rely on the model to estimate future instances.

Listing 2-12. Ordinary Least-Squares Regression Model Development Applying Statsmodels

```
import statsmodels.api as sm
from sklearn.model_selection import train_test_split
from sklearn.preprocessing import StandardScaler
x = np.array(df["lending_rate"])
y = np.array(df["final_consumption"])
x = x.reshape(-1,1)
y = y.reshape(-1,1)
x_train, x_test, y_train, y_test = train_test_split(x,y,test_size=0.2,
shuffle=False)
scaler = StandardScaler()
x_train = scaler.fit_transform(x_train)
x_test = scaler.transform(x_test)
x_constant = sm.add_constant(x_train)
x_test = sm.add_constant(x_test)
model = sm.OLS(y_train,x_constant).fit()
model.summary()
```

Table 2-7. *Ordinary Least-Squares Regression Model Results*

Dep. Variable	y	R-squared	0.443
Model	OLS	Adj. R-squared	0.431
Method	Least Squares	F-Statistic	36.64
Date	Wed, 04 Aug 2021	Prob (F-statistic)	2.41e-07
Time	11:06:05	Log-Likelihood	-1454.2
No. Observations	48	AIC	2912.
Df Residual	46	BIC	2916.
Df Model	1		
Covariance Type:	nonrobust		

(*continued*)

Table 2-7. (*continued*)

	coef	std err	t	P>ltl	[0.025	0.975]
const	7.427e+12	5.12e+11	14.496	0.000	6.4e+12	8.46e+12
x1	-3.102e+12	5.12e+11	-6.053	0.000	-4.13e+12	-2.07e+12

Omnibus:	1.135	**Durbin-Watson:**	1.754
Prob(Omnibus):	0.567	Jarque-Bera (JB):	1.111
Skew:	0.236	Prob(JB):	0.574
Kurtosis:	2.424	Cond. No.	1.00

Table 2-7 shows that the ordinary least-squares regression model explains 53.2% of the variability in the data (the R^2 score is 0.443). Using findings from this table, Equation 2-8 shows how changes in the U.S. lending interest rate influence changes in the final U.S. consumption expenditure.

$$U.S.FCE = \beta_0 - 3.102e + 12\big(Lending\ Interest\ Rate(\%)\big) + \varepsilon_i \quad \text{(Equation 2-8)}$$

Equation 2-8 states that, for each unit change in the U.S. lending interest rate, the final consumption expenditure decreases by 3.102e+12. In addition, Table 2-7 reveals that the mean value of the predicted U.S. final consumption expenditure—when you hold the U.S. lending interest rate constant—is 7.427e+12. Last, the table tells you that the model explains about 36% of the variability in the data—it struggles to predict future instances of the U.S. final consumption expenditure.

Ordinary Least-Squares Regression Model Development Using Scikit-Learn

The preceding section revealed a way to develop and evaluate the ordinary least-squares model using the statsmodels library. Findings show that the expressed macroeconomic phenomenon does indeed exist. You cannot fully rely on the model to predict future instances of the U.S. final consumption expenditure, given the poor performance obtained in the test.

This section studies the same phenomenon, but applies a different machine learning library to develop and assess the model. It uses a popular open source library called scikit-learn. Developing this model is similar to the way it was done in the previous section. This approach has a competitive edge over statsmodels because it provides an easier way to control how the model behaves and to validate its performance. In addition, it has a wide range of regression models, i.e., Ridge, Lasso, and ElasticNet, among others. This chapter does not cover those models. All you need to know for now is that these models extend the ordinary least-squares model by introducing a term that penalizes the model for errors it makes in the test. The primary difference between statsmodels and sklearn is that statsmodels is better at hypothesis testing and sklearn is better at predictions and does not return t-statistics. See Listing 2-13.

Listing 2-13. Ordinary Least-Squares Regression Model Development Applying Scikit-Learn

```
x = np.array(df["lending_rate"])
y = np.array(df["final_consumption"])
x = x.reshape(-1,1)
y = y.reshape(-1,1)
x_train, x_test, y_train, y_test = train_test_split(x,y,test_size=0.2,
random_state=0)
scaler = StandardScaler()
x_train = scaler.fit_transform(x_train)
x_test = scaler.transform(x_test)
from sklearn.linear_model import LinearRegression
lm = LinearRegression()
lm.fit(x_train,y_train)
```

Cross-Validation

Listing 2-14 applies the cross_val_score() method to validate the performance of the default ordinary least-squares regression model over different subsets of the same data. It applies R^2 to find a score, since sklearn does not calculate the adjusted R^2. It then estimates the mean and standard deviation of the validation score.

Listing 2-14. Ordinary Least-Squares Regression Model Cross-Validation

```
from sklearn.model_selection import cross_val_score
def get_val_score(scores):
    lm_scores = cross_val_score(scores, x_train, y_train, scoring="r2")
    print("CV mean: ", np.mean(scores))
    print("CV std: ", np.std(scores))
get_val_score(lm)
CV mean:   0.2424291076433461
CV std:   0.11822093588298
```

The results show that the mean data point of the cross-validation score is 0.2424 and independent data points of the cross-validation score deviate from the mean data point by 0.1182. Listing 2-15 shows where the data spill begins in the data, so that you can create a dataframe that uses actual data points and the predicted data points with a date index.

Listing 2-15. Find Where the Spill Begins in the Data

```
y_test.shape
(13, 1)
```

Predictions

Listing 2-16 tabulates the actual data points and the predicted data points of the U.S. final consumption expenditures side-by-side (see Table 2-8).

Listing 2-16. The Actual U.S. Final Consumption Expenditure and Predicted U.S. Final Consumption Expenditures

```
actual_values = pd.DataFrame(y_test)
actual_values.columns = ["Actual FCE"]
predicted_values  = pd.DataFrame(lm.predict(x_test))
predicted_values.columns = ["Predicted FCE"]
actual_and_predicted_values = pd.concat([actual_values,
predicted_values],axis=1)
actual_and_predicted_values.index = pd.to_datetime(df[:13].index)
actual_and_predicted_values
```

Table 2-8. *The Actual and Predicted U.S. Final Consumption Expenditures*

Date	Actual FCE	Predicted FCE
2020-01-01	5.829066e+12	6.511152e+12
2019-01-01	3.403470e+12	3.385294e+12
2018-01-01	7.173358e+12	9.102986e+12
2017-01-01	5.245942e+12	7.502778e+12
2016-01-01	1.227281e+13	1.085977e+13
2015-01-01	1.688457e+13	9.009788e+12
2014-01-01	3.624267e+12	5.175626e+12
2013-01-01	7.173358e+12	9.461798e+12
2012-01-01	2.204317e+12	6.560470e+12
2011-01-01	7.144856e+12	5.151394e+12
2010-01-01	1.543082e+13	1.056713e+13
2009-01-01	1.269598e+13	1.085977e+13
2008-01-01	4.757180e+12	3.300484e+12

To plot the data points in Table 2-8, use Listing 2-17 (see Figure 2-9).

Listing 2-17. The Actual and Predicted U.S. Final Consumption Expenditures

```
actual_and_predicted_values.plot(lw=4)
plt.xlabel("Date")
plt.title("The U.S. actual and predicted FCE")
plt.ylabel("FCE")
plt.legend(loc="best")
plt.show()
```

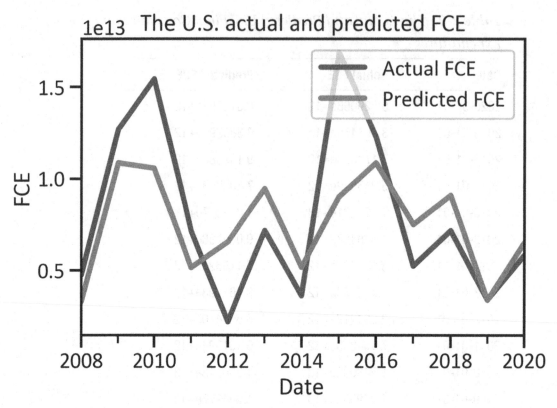

Figure 2-9. *The actual and predicted U.S. final consumption expenditures*

Figure 2-9 demonstrates the actual data points and predicted data points of the U.S. final consumption expenditure. It shows how the predicted data points deviate from the actual data points.

Estimating Intercept and Coefficients

Listing 2-18 returns estimates of the coefficient (the difference between the predicted data point of the U.S. final consumption expenditure for each one-unit change in the U.S. lending interest rate). Note the definition applies only to linear models. Equation 2-9 expresses the regression equation for this chapter.

Listing 2-18. Estimate Intercept

```
lm.intercept_
array([6.95280973e+12])
```

The mean value of the predicted U.S. final consumption expenditure—when you hold the U.S. lending interest rate constant—is 7.56456705e+12. Listing 2-19 estimates the coefficient.

Listing 2-19. Estimate Coefficient

```
lm.coef_
array([[-2.71748602e+12]])
```

Equation 2-9 expresses the regression equation for this chapter.

$$U.S.FCE = 6.95280973e - 2.71748602e + 12(U.S.lending\ rate) + \varepsilon_i \qquad \text{(Equation 2-9)}$$

Equation 2-9 highlights that, for each unit change in the U.S. lending interest rate, the U.S. final consumption expenditure decreases by 2.71748602e+122.

Figure 2-10 plots the data fed into the ordinary least-squares regression model.

Listing 2-20. The U.S. Lending Interest Rate and Final Consumption Expenditure

```
plt.scatter(x_train,y_train,color="navy",s=200)
plt.title("The U.S. lending interest rate (%) and FCE")
plt.ylabel("Actual U.S. FCE")
plt.xlabel("Actual U.S. lending interest rate (%)")
plt.show()
```

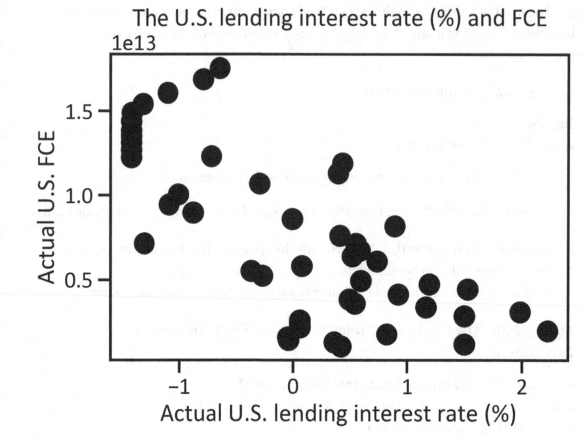

Figure 2-10. *The U.S. lending interest rate and the final consumption*
expenditure

Figure 2-10 plots the actual data points of the U.S. lending interest rate (%) and the predicted data points of the final consumption expenditure.

Listing 2-21. The U.S. Final Consumption Expenditure Predictions

```
plt.scatter(x_test,predicted_values,color="navy",s=200)
plt.plot(x_test,predicted_values,color="red",lw=4)
plt.title("The U.S. FCE predictions")
plt.xlabel("Actual U.S. lending interest rate (%)")
plt.ylabel("Predicted U.S. FCE")
plt.show()
```

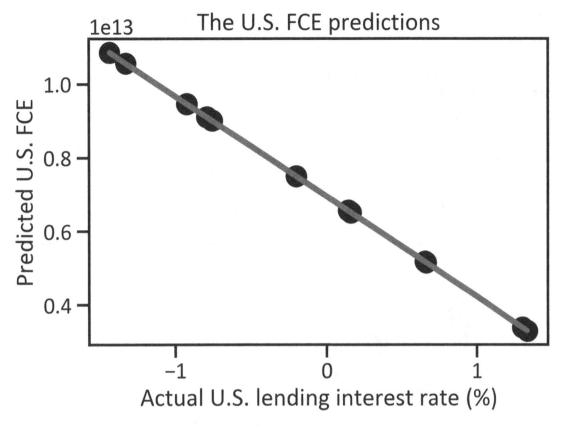

Figure 2-11. *The U.S. final consumption expenditure predictions*

Figure 2-11 shows that the actual data points of the U.S. lending interest rate and the predicted data points of the final consumption expenditure are perfectly linear, by construction.

Residual Analysis

A *residual* represents an estimate of the difference between the data points that the linear regression method estimates and the actual data points. Equation 2-10 defines residuals.

$$\varepsilon_i = y - \hat{y} \qquad \text{(Equation 2-10)}$$

The ordinary least-squares regression model comes with the normality assumption.

Listing 2-22. The U.S. Actual and Predicted Values of Final Consumption
Expenditures

```
plt.scatter(y_test,predicted_values,color="navy",s=200)
plt.title("The U.S. actual and predicted FCE")
plt.xlabel("Actual U.S. FCE")
plt.ylabel("Predicted U.S. FCE")
plt.show()
```

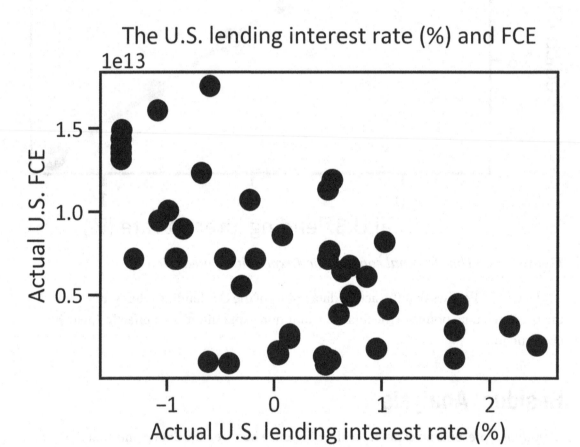

Figure 2-12. *The U.S. actual and predicted values of the final consumption*
expenditures

Figure 2-12 shows that the actual data points and the predicted data points of the
U.S. final consumption expenditure are randomly distributed. Figure 2-13 shows the
central tendency and the dispersion of residuals.

Listing 2-23. Ordinary Least-Squares Regression Model Residual Distribution

```
residuals = y_test - predicted_values
residuals = pd.DataFrame(residuals)
residuals.columns = ["Residuals"]
residuals.plot(kind="box",color="navy")
plt.title("Model residual distribution")
plt.ylabel("Values")
plt.show()
```

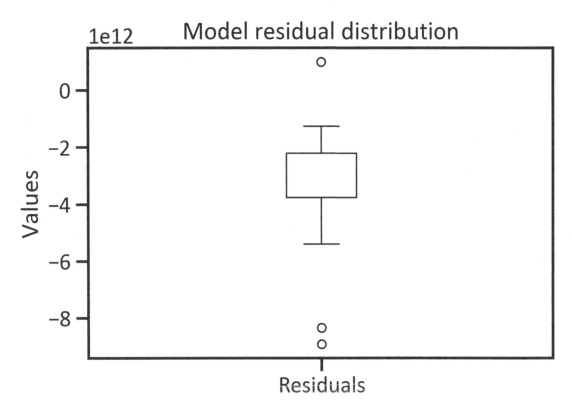

Figure 2-13. *Ordinary least-squares regression model residual distribution*

Figure 2-13 shows that the mean data point is far from 0. This indicates that the ordinary least-squares regression model does not meet the assumption. Figure 2-14 shows the serial correlation. It shows the number of *lags* (how far back the residuals in the present time are related to themselves) on the x-axis and the autocorrelation in the y-axis. Straight lines with blue dots are the autocorrelation coefficients. If there are any autocorrelation correlation coefficients that are outside the statistical control (the blue region), the model does not meet the regression assumptions.

Listing 2-24. Ordinary Least-Squares Regression Model Residual Autocorrelation

```
from statsmodels.graphics.tsaplots import plot_acf
plot_acf(residuals)
plt.title("Model residuals autocorrelation")
plt.xlabel("Lags")
plt.ylabel("ACF")
plt.show()
```

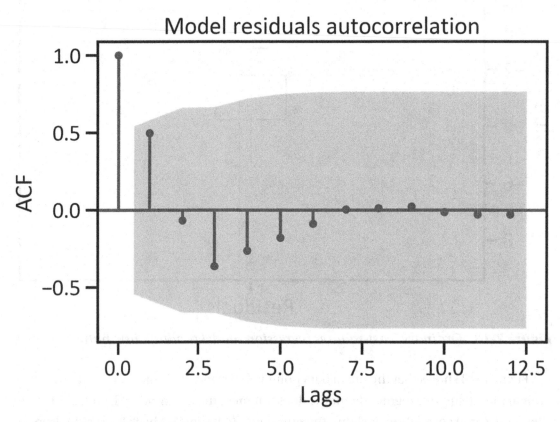

Figure 2-14. *Ordinary least-squares regression model residuals autocorrelation*

Figure 2-14 demonstrates that after the lag, there are no autocorrelation coefficients that are outside the blue region (the confidence interval).

Other Ordinary Least-Squares Regression Model Performance Metrics

Other ordinary least-squares regression model performance metrics include the following:

- **Mean Absolute Error**—Represents the average magnitude of the errors prior to regressing the variables.

- **Mean Squared Error**—Represents an average of the sum of errors.

- **Root-MSE**—Represents the variability explained after regressing variables. Apply RMSE when you do not want huge residuals.

- **R-squared**—Represents the extent to which a regressor explains the variability in the data.

Listing 2-25. Ordinary Least-Squares Regression Model Performance Matrix

```
from sklearn import metrics
MAE = metrics.mean_absolute_error(y_test,predicted_values)
MSE = metrics.mean_squared_error(y_test,predicted_values)
RMSE = np.sqrt(MSE)
EV = metrics.explained_variance_score(y_test,predicted_values)
R2 = metrics.r2_score(y_test,predicted_values)
lmmodelevation = [[MAE,MSE,RMSE,EV,R2]]
lmmodelevationdata = pd.DataFrame(lmmodelevation,
index=["Values"],columns=("Mean absolute error", "Mean squared error",
"Root mean squared error", "Explained variance score", "R-squared"))
lmmodelevationdata
```

Table 2-9. *Ordinary Least-Squares Regression Model Performance Matrix*

	Mean Absolute Error	Mean Squared Error	Root Mean Squared Error	Explained Variance Score	R-Squared
Values	2.501582e+12	1.023339e+25	3.198966e+12	0.524619	0.513118

Table 2-9 shows that the ordinary least-squares regression model explains 51.31% of the variability in the data. The magnitude average of the errors prior to regressing variables is 2.501582e+12 and the average sum of error is 1.023339e+25. Last, the explained variance score is 52.46%, further confirming mediocrity in the model.

Ordinary Least-Squares Regression Model Learning Curve

Figure 2-15 shows how the ordinary least-squares model learns to explain the variability in the relationship between the U.S. lending interest rate and the final U.S. consumption expenditure as the train test size increases. In summary, it shows how the R^2 score changes as you increase the train test size. See Listing 2-26.

Listing 2-26. Ordinary Least-Squares Regression Model Learning Curve

```
from sklearn.model_selection import learning_curve
trainsizelogreg, trainscorelogreg, testscorelogreg = learning_curve(lm,
x,y,cv=5, n_jobs=-1, scoring="r2", train_sizes=np.linspace(0.1,1.0,50))
trainscorelogreg_mean = np.mean(trainscorelogreg,axis=1)
trainscorelogreg_std = np.std(trainscorelogreg,axis=1)
testscorelogreg_mean = np.mean(testscorelogreg,axis=1)
testscorelogreg_std = np.std(testscorelogreg,axis=1)
fig, ax = plt.subplots()
plt.plot(trainsizelogreg,trainscorelogreg_mean, label="Training R2 score",
color="red",lw=4)
plt.plot(trainsizelogreg,testscorelogreg_mean, label="Validation R2
score",color="navy", lw=4)
plt.title("OLS learning curve")
plt.xlabel("Training set size")
```

```
plt.ylabel("R2 score")
plt.legend(loc=4)
plt.show()
```

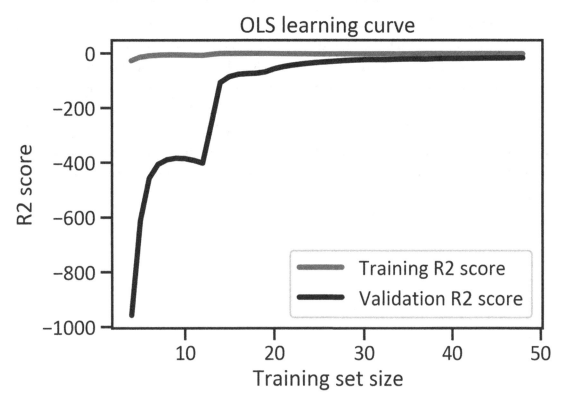

Figure 2-15. *Ordinary least-squares regression model learning curve*

Figure 2-15 demonstrates that the cross-validation R^2 score is lower than the training R^2 score throughout the training process—this indicates that the R^2 score is high in training when you carry out repeated tests. During the validation stage, as the ordinary least-squares regression model approaches the tenth data point in the data, it begins to encounter difficulties predicting data points of the U.S. final consumption expenditure. Meanwhile, during training, the regressor consistently predicts the data points.

Conclusion

This chapter introduced a way of finding answers to cross-sectional data by applying the ordinary least-squares regression model. It first determined the spread of data points, then handled any missing data points and outliers. Then it examined the joint variability and association between the U.S. lending interest rate (as a percentage) and the final consumption expenditure (in current U.S. dollars). Next, it applied the Pearson correlation method and found that there was negative correlation between the U.S. lending interest rate and the final U.S. consumption expenditure.

Following that, the chapter showed you how to develop the ordinary least-squares regression model using the `statsmodels` and `scikit-learn` libraries. The model developed using `statsmodels` showed superior performance compared to the one developed using `scikit-learn`.

CHAPTER 3

Multivariate Consumption Study Applying Regression

The last chapter covered multivariate linear regression—a model for predicting continuous response variables by applying a single predictor variable. There will be cases where you'll have more than one predictor variable, so this chapter presents ways of properly fitting multiple variables into a regression equation. It applies the ordinary least-squares regression model to determine whether changes in the U.S. social contributions (the current LCU), lending interest rate (as a percentage), and GDP (Gross Domestic Product) growth as an annual percentage influence changes in the final consumption expenditure (in current U.S. dollars).

To begin, this model checks whether the data comes from a normal distribution. Following that, it applies the Pearson correlation method to investigate the correlation among the variables, and then it implements the Eigen matrix to determine the severity among the variables. It then reduces the data into two dimensions by applying a dimension reduction technique called *principal component analysis*. It then fits the predictor variables into the regression model to predict the U.S. final consumption expenditure. Finally, it determines the central tendency and distribution of residuals and investigates how the model learns to make reliable predictions as you increase training data.

© Tshepo Chris Nokeri 2022
T. C. Nokeri, *Econometrics and Data Science*, https://doi.org/10.1007/978-1-4842-7434-7_3

Context of This Chapter

Predictor variables include the U.S. social contributions (current LCU), lending interest rate (as a percentage), and GDP growth (as an annual percentage) and the response variable is the U.S. final consumption expenditure (in current U.S. dollars). Table 3-1 outlines macroeconomic indicators and shows their code.

Table 3-1. The U.S. Macroeconomic Indicators for This Chapter

Code	Title
GC.REV.SOCL.CN	The U.S. social contributions (current LCU)
FR.INR.LEND	The U.S. lending interest rate (as a percentage)
NY.GDP.MKTP.KD.ZG	The U.S. GDP growth (as an annual percentage)
NE.CON.TOTL.CD	The U.S. final consumption expenditure (in current U.S. dollars)

Social Contributions (Current LCU)

Social contributions represent social security contributions by both organizations and employees (see Listing 3-1). This might additionally include contributions that the government adds for insurance. Figure 3-1 illustrates the U.S. social contributions from 1971 to 2021.

Listing 3-1. The U.S. Social Contributions (Current LCU)

```
import wbdata
country = ["USA"]
indicator = {"GC.REV.SOCL.CN":"social_contri"}
social_contri = wbdata.get_dataframe(indicator, country=country,
convert_date=True)
social_contri.plot(kind="line",color="green",lw=4)
plt.title("The U.S. social contributions")
plt.ylabel("Social contributions")
plt.xlabel("Date")
plt.show()
```

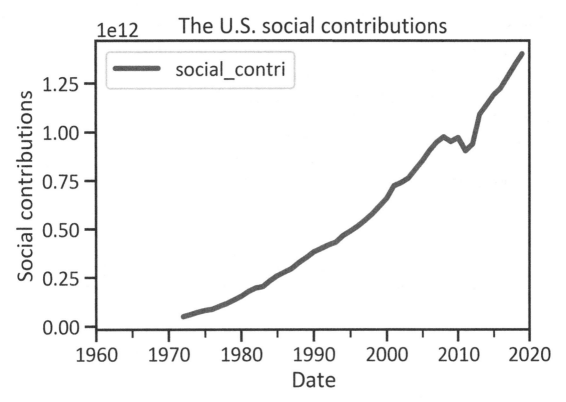

Figure 3-1. *The U.S. social contributions (current LCU)*

Figure 3-1 shows an unmistakable upward trend in the U.S. social contribution data from 1971 to 2020. This shows that U.S.-based organizations and employees currently contribute to social security more than ever. Note that this information does not account for inflation—the indicator does not fully capture the change in those contributions.

Lending Interest Rate

A lending interest rate represents a rate at which financial institutions charge their customers when providing them with loans (see Listing 3-2). It reflects the cost of borrowing money. A low lending interest rate indicates that borrowers pay less when servicing debt. In contrast, a high interest rate indicates that the cost of borrowing money is high. Figure 3-2 illustrates the U.S. lending interest rate from 1971 to 2020.

Listing 3-2. The U.S. Lending Interest Rate

```
country = ["USA"]
indicator = {"FR.INR.LEND":"lending_rate"}
lending_rate = wbdata.get_dataframe(indicator, country=country,
convert_date=True)
social_contri.plot(kind="line",color="orange",lw=4)
plt.title("The U.S. lending interest rate (%)")
plt.ylabel("Lending interest rate (%)")
plt.xlabel("Date")
plt.show()
```

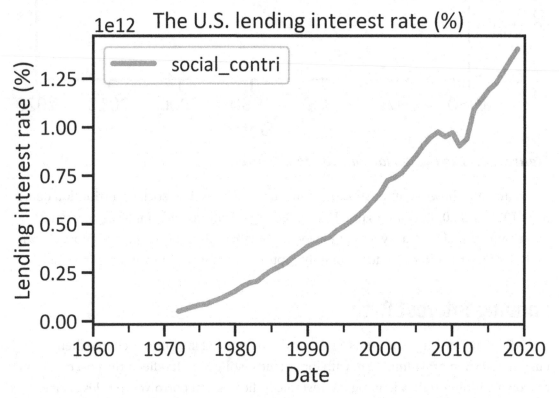

Figure 3-2. *The U.S. lending interest rate*

Figure 3-2 shows that the U.S. lending interest rate has grown by over 1.25% in the last five decades.

GDP Growth (Annual Percentage)

GDP stands for Gross Domestic Product. It estimates the market value of finished goods and services that households of a country consume annually. It is an exemplary indicator for diagnosing the health of an economy, as it conveys the country's economic production. There are two main ways of estimating GDP, using the *nominal* GDP (a GDP estimate that does not consider inflation), and using the *real* GDP, which is a GDP estimate that considers inflation. Equation 3-1 is the GDP formula.

$$Y = C + I + G + (X - M)$$
(Equation 3-1)

Where Y represents GDP, C represents consumer spending, I represent investments, G represents government spending, X represents exports, and M represents imports. Figure 3-3 shows the changes in annual percentage growth of the GDP in the U.S. from 1971 to 2020. See Listing 3-3.

Listing 3-3. The U.S. GDP Growth (Annual %)

```
country = ["USA"]
gdp_growth = {"NY.GDP.MKTP.KD.ZG":"gdp_growth"}
gdp_growth = wbdata.get_dataframe(indicator, country=country,
convert_date=True)
gdp_growth.plot(kind="line",color="navy",lw=4)
plt.title("The U.S. GDP growth (annual %)")
plt.ylabel("GDP growth (annual %)")
plt.xlabel("Date")
plt.show()
```

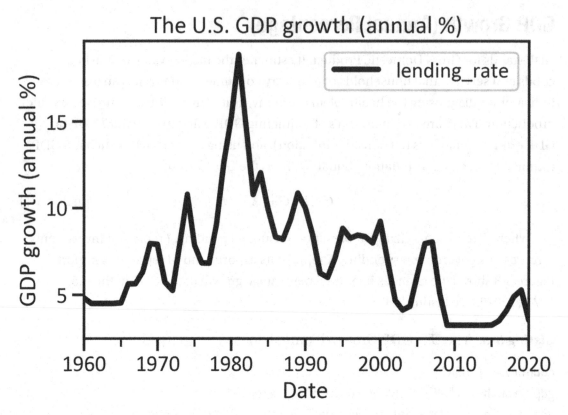

Figure 3-3. *The U.S. GDP growth*

Figure 3-3 shows the zig zag pattern of the U.S. GDP, which has been adjusted for inflation. In 1981, the United States experienced its highest change in GDP. In 2019, on the other hand, economic activity was very low (the GDP grew by 2.161176).

Final Consumption Expenditure

The final consumption expenditure is the market value of all goods and services purchased by households (see Listing 3-4). This includes durable products such as cars and personal computers. Figure 3-4 illustrates the U.S. final consumption expenditure in current U.S. dollars.

Listing 3-4. The U.S. Final Consumption Expenditure (Current U.S. Dollars)

```
country = ["USA"]
indicator = {"NE.CON.TOTL.CD":"fin_cons"}
fin_cons = wbdata.get_dataframe(indicator, country=country,
convert_date=True)
social_contri.plot(kind="line",color="red",lw=4)
plt.title("The U.S. FCE")
plt.ylabel("FCE")
plt.xlabel("Date")
plt.show()
```

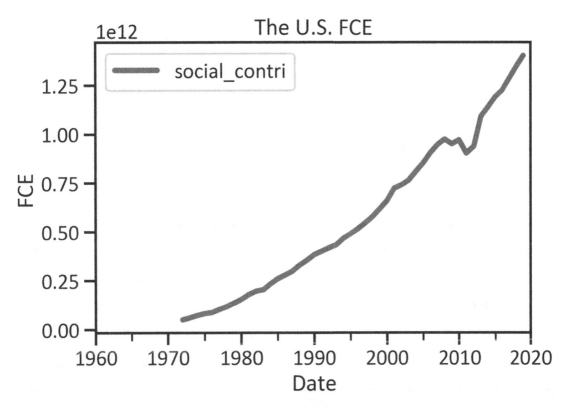

Figure 3-4. *The U.S. final consumption expenditure*

Figure 3-4 shows an extreme upward trend in the U.S. final consumption expenditure data, which was slightly disrupted in 2012. Next, you explore a theoretical framework in order to have some background on the phenomenon.

Theoretical Framework

This chapter applies the multivariate regression method to determine whether a set of predictor variables (the U.S. social contributions, the lending interest rate, and the GDP growth) influence the response variable (the U.S. final consumption expenditure). Figure 3-5 shows the theoretical framework of the problem at hand.

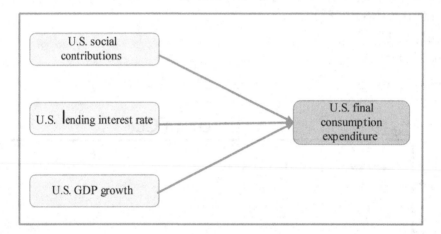

Figure 3-5. *Theoretical framework*

Based on Figure 3-5, the chapter tests the following hypotheses.

HYPOTHESES 1

H_0: There is no significant difference between the U.S. social contributions and the final consumption expenditure.

H_A: There is a significant difference between the U.S. social contributions and the final consumption expenditure.

HYPOTHESES 2

H_0: There is no significant difference between the U.S. lending interest rate and the final consumption expenditure.

H_A: There is a significant difference between the U.S. lending interest rate and the final consumption expenditure.

HYPOTHESES 3

H_0: There is no significant difference between the U.S. GDP growth and the final consumption expenditure.

H_A: There is a significant difference between the U.S. GDP growth and the final consumption expenditure.

This chapter applies the ordinary least-squares regression model to investigate whether changes in the following three affect the consumption patterns in U.S. households:

- The rate financial institutions in the U.S. charge customers for lending

- The annual growth in the U.S. economic activities considering inflation

- Contributions that U.S. organizations and citizens make to social security

Listing 3-5 applies wbdata to acquire data associated with the U.S. social contributions, the lending interest rate, the GDP growth, and final consumption expenditure (see Table 3-2).

Listing 3-5. Load the U.S. Macroeconomic Data

```
country = ["USA"]
indicator = {"GC.REV.SOCL.CN":"social_contri",
             "FR.INR.LEND":"lending_rate",
             "NY.GDP.MKTP.KD.ZG":"gdp_growth",
             "NE.CON.TOTL.CD":"fin_cons"}
df = wbdata.get_dataframe(indicator, country=country, convert_date=True)
df.head()
```

Table 3-2. *Load the U.S. Macroeconomic Data*

Date	social_contri	lending_rate	gdp_growth	fin_cons
2020-01-01	1.420776e+12	3.544167	-3.486140	NaN
2019-01-01	1.402228e+12	5.282500	2.161177	1.753966e+13
2018-01-01	1.344612e+12	4.904167	2.996464	1.688457e+13
2017-01-01	1.283672e+12	4.096667	2.332679	1.608306e+13
2016-01-01	1.224603e+12	3.511667	1.711427	1.543082e+13

Descriptive Statistics

Most machine learning models are exceedingly sensitive to missing data points (blank data points in a certain column or row). Consequently, you have to replace them with a central value (the mean or median value). When applying the pandas library, you can use the fillna() method with a central value that must be used for each missing value in a certain column. Listing 3-6 substitutes the missing data points with the mean value and Listing 3-7 shows the U.S. social contributions distribution.

Listing 3-6. Substitute Missing with the Mean Value

```
df["social_contri"] = df["social_contri"].fillna(df["social_contri"].mean())
df["lending_rate"] = df["lending_rate"].fillna(df["lending_rate"].mean())
df["gdp_growth"] = df["gdp_growth"].fillna(df["gdp_growth"].mean())
df["fin_cons"] = df["fin_cons"].fillna(df["fin_cons"].mean())
```

Figure 3-6 shows the distribution of the U.S. social contributions.

Listing 3-7. The U.S. Social Contributions Distribution

```
df["social_contri"] .plot(kind="box",color="green")
plt.title("The U.S. social contributions' distribution")
plt.ylabel("Values")
plt.show()
```

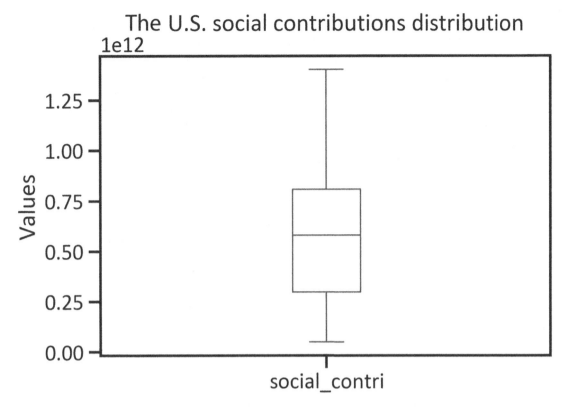

Figure 3-6. *The U.S. social contributions distribution*

Figure 3-6 shows that the U.S. social contribution data possess no outliers. Figure 3-7 shows the U.S. lending interest rate. See Listing 3-8.

Listing 3-8. The U.S. Lending Interest Rate Distribution

```
df["lending_rate"] .plot(kind="box",color="orange")
plt.title("US lending interest rate (%) distribution")
plt.ylabel("Values")
plt.show()
```

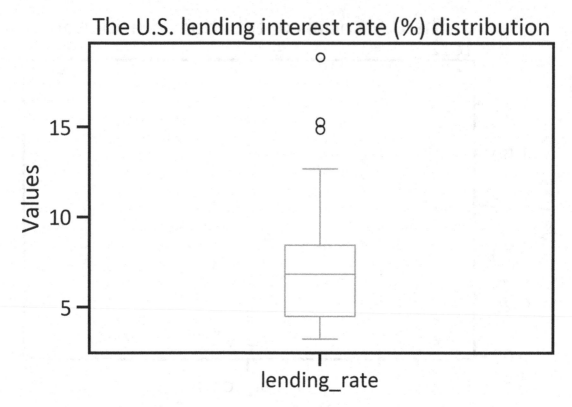

Figure 3-7. *The U.S. lending interest rate distribution*

Figure 3-7 demonstrates that there are three outliers in the U.S. lending interest rate data. Listing 3-9 substitutes these outliers with the mean value. Listing 3-10 shows the GDP growth distribution.

Listing 3-9. Replacing U.S. Lending Interest Rate Outliers with the Mean Value

```
df['lending_rate'] = np.where((df["lending_rate"] > 13),df["lending_rate"].
mean(),df["lending_rate"])
```

Figure 3-8 demonstrates the distribution of U.S. GDP growth.

Listing 3-10. The U.S. GDP Growth Distribution

```
df["gdp_growth"] .plot(kind="box",color="navy")
plt.title("The U.S. GDP growth (annual %) distribution")
plt.ylabel("Values")
plt.show()
```

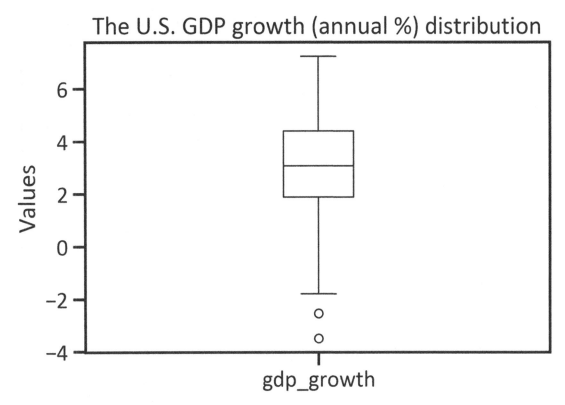

Figure 3-8. *The U.S. GDP growth distribution*

Figure 3-8 demonstrates that there are two outliers in the U.S. GDP growth data. Listing 3-11 substitutes these outliers with the mean value. Listing 3-12 shows the final consumption expenditure.

Listing 3-11. Replacing U.S. GDP Growth Outliers with the Mean Value

```
df['gdp_growth'] = np.where((df["gdp_growth"] < 0),df["gdp_growth"].
mean(),df["gdp_growth"])
```

Figure 3-9 shows U.S. final consumption expenditure (in current U.S. dollars).

Listing 3-12. The U.S. Final Consumption Expenditure

```
df["fin_cons"] .plot(kind="box",color="red")
plt.title("The U.S. FCE distribution")
plt.ylabel("Values")
plt.show()
```

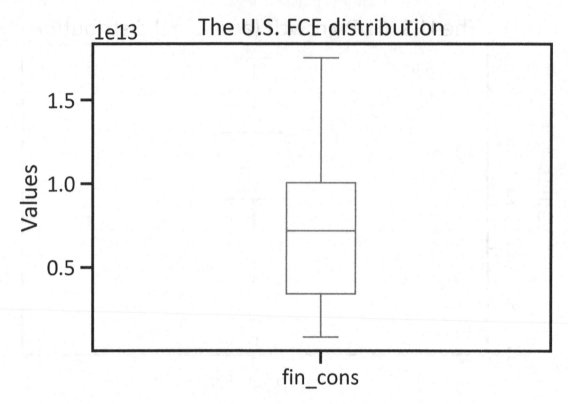

Figure 3-9. *The U.S. final consumption expenditure*

Figure 3-9 demonstrates that there are no outliers in this data. Table 3-3 outlines the central tendency and dispersion in the data, which you can view using the command in Listing 3-13.

Listing 3-13. Descriptive Summary

```
df.describe().transpose()
```

Table 3-3. Descriptive Summary

	Count	Mean	Std	Min	25%	50%	75%	Max
social_contri	61.0	5.977005e+11	3.687861e+11	5.048000e+10	2.985900e+11	5.977005e+11	8.533970e+11	1.420776e+12
lending_rate	61.0	6.639908e+00	2.436502e+00	3.250000e+00	4.500000e+00	6.824167e+00	8.270833e+00	1.266583e+01
gdp_growth	61.0	3.516580e+00	1.349463e+00	9.983408e-01	2.684287e+00	3.075515e+00	4.400000e+00	7.236620e+00
fin_cons	61.0	7.173358e+12	4.580756e+12	8.395100e+11	3.403470e+12	7.173358e+12	1.006535e+13	1.753966e+13

Table 3-3 shows that:

- The mean value of the U.S. social contributions is 5.977005e+11.

- The lending interest rate is 6.639908e+00.

- The GDP growth is 3.516580e+00.

- The final consumption expenditure is 7.173358e+12.

- The U.S. social contributions deviate from the mean value by 3.687861e+11, the lending interest rate deviates by 2.436502e+00, the GDP growth deviates by 1.349463e+00, and the final consumption expenditure deviates by 4.580756e+12.

Covariance Analysis

Listing 3-14 determines how the variables co-vary and how each variable varies with itself (see Table 3-4).

Listing 3-14. Covariance Matrix

```
dfcov = df.cov()
dfcov
```

Table 3-4. *Covariance Matrix*

	social_contri	lending_rate	gdp_growth	fin_cons
social_contri	1.360032e+23	-6.000783e+11	-2.202567e+11	1.553786e+24
lending_rate	-6.000783e+11	5.936544e+00	8.538603e-01	-7.092189e+12
gdp_growth	-2.202567e+11	8.538603e-01	1.821052e+00	-2.659879e+12
fin_cons	1.553786e+24	-7.092189e+12	-2.659879e+12	2.098332e+25

Table 3-4 outlines the estimated covariance of the set of variables you retrieved.

Correlation Analysis

Listing 3-15 applies the previous covariance estimates and divides them into the standard deviation of the predictor variable and into the response variable to find the Pearson correlation coefficients (see Figure 3-10).

Listing 3-15. Pearson Correlation Matrix

```
dfcorr = df.corr(method="pearson")
sns.heatmap(dfcorr, annot=True, annot_kws={"size":12},cmap="Blues")
plt.title("Pearson correlation matrix")
plt.show()
```

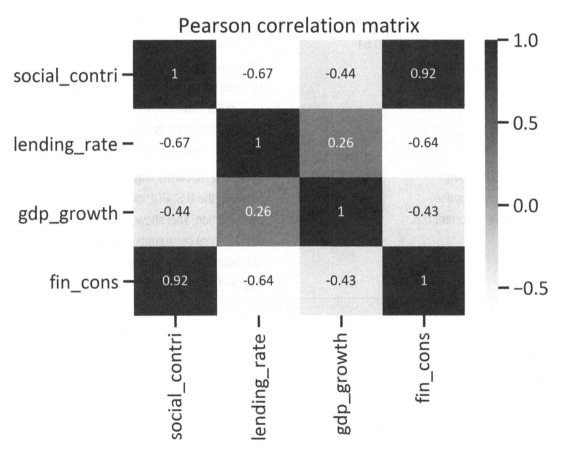

Figure 3-10. *Pearson correlation matrix*

Table 3-5 interprets the Pearson correlation coefficients in Figure 3-10.

Table 3-5. *Interpretation of Pearson Correlation Coefficients*

Variables	Pearson Correlation Coefficient	Findings
The U.S. social contributions (current LCU) and final consumptions expenditure (in current U.S. dollars)	0.92	There is an extreme positive correlation between the U.S. social contributions and the final consumption. This shows that on average, the U.S. final consumption expenditure tends to sharply increase when U.S. organizations and individuals increase their contributions to social security.
The U.S. lending interest rate and final consumption expenditure (in current U.S. dollars)	-0.64	There is a moderate negative correlation between the U.S. lending interest rate and the final consumption. This shows that on average, the U.S. final consumption expenditure tends to moderately decline as the cost of borrowing goes up.
The U.S. GDP growth and final consumption expenditure in current U.S. dollars	-0.43	There is a moderate negative correlation between the U.S. GDP growth and the final consumption. This shows that on average, the U.S. final consumption expenditure tends to moderately decrease as economic activity, adjusted for inflation, increases.

Figure 3-11 shows the distribution of the U.S. social contributions, lending interest rate, and the final consumption expenditure, including the statistical dependence among these variables. Use the command in Listing 3-16 to view these pair plots.

Listing 3-16. Pair Plot

```
sns.pairplot(df)
```

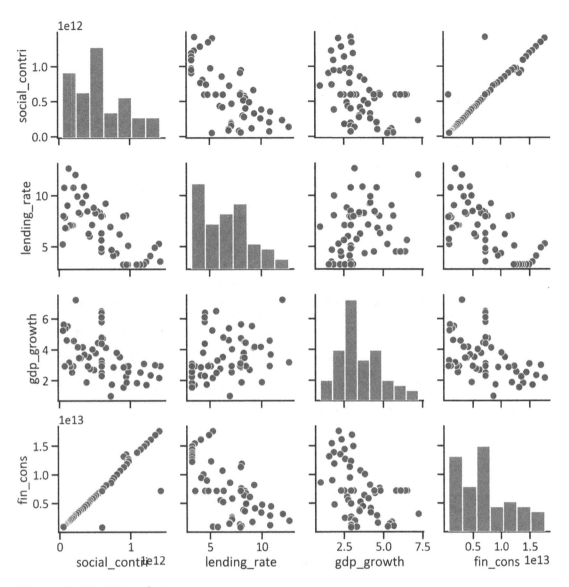

Figure 3-11. *Pair plots*

Correlation Severity Detection

Listing 3-17 uses the Eigen method to detect the correlation severity. It incorporates *eigenvalues*, which are vectors we consider after a linear transform, and eigenvectors, also called Eigen loadings (see Table 3-6).

Listing 3-17. Determine Correlation Severity

```
eigenvalues, eigenvectors = np.linalg.eig(dfcorr)
eigenvalues = pd.DataFrame(eigenvalues)
eigenvalues.columns = ["Eigen values"]
eigenvectors = pd.DataFrame(eigenvectors)
eigenvectors.columns = df.columns
eigenvectors = pd.DataFrame(eigenvectors)
eigenmatrix = pd.concat([eigenvalues, eigenvectors],axis=1)
eigenmatrix.index = df.columns
eigenmatrix
```

Table 3-6. *Eigen Matrix*

	Eigen Values	social_contri	lending_rate	gdp_growth	fin_cons
social_contri	2.741814	0.571562	-0.727660	-0.363008	-0.109782
lending_rate	0.078913	-0.477992	-0.051401	-0.774123	0.411844
gdp_growth	0.403418	-0.355849	-0.015693	-0.257158	-0.898329
fin_cons	0.775854	0.564103	0.683829	-0.450365	-0.106477

Table 3-7 explains how to interpret the Eigen matrix.

Table 3-7. *Eigen Matrix Interpretation*

Eigenvalue Score	Description
> 100	There is extreme multicollinearity.
10 - 100	There is moderate multicollinearity.
0 - 10	There is slight multicollinearity.

Based on Table 3-6, there is multicollinearity in the data. Therefore, you need to perform a dimension reduction.

Dimension Reduction

It is time-consuming to individually investigate each relationship among the variables when there are many of them. Listing 3-18 reduces the data and Figure 3-12 demonstrates the variance ratio of each component.

Listing 3-18. Dimension Reduction

```
from sklearn.decomposition import PCA
from sklearn.preprocessing import StandardScaler
scaler = StandardScaler()
std_df = scaler.fit_transform(df)
pca = PCA()
pca.fit_transform(std_df)
pca_variance = pca.explained_variance_
plt.figure(figsize=(8, 6))
plt.bar(range(4), pca_variance, align="center", label="Individual
variance")
plt.legend()
plt.ylabel("Variance ratio")
plt.xlabel("Principal components")
plt.title("Individual variance")
plt.show()
```

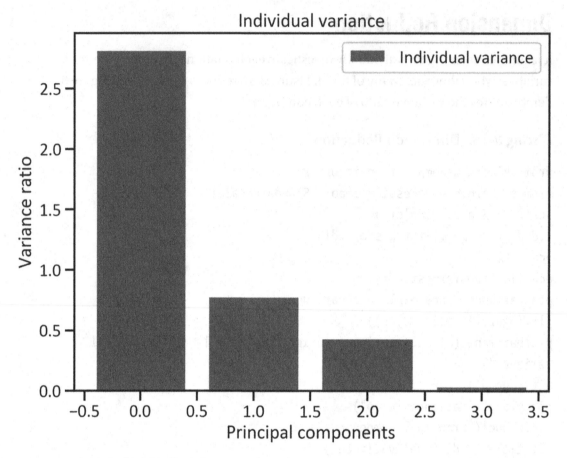

Figure 3-12. *Individual variance*

Figure 3-12 shows that the variance of the first component is over 2.5. Listing 3-19 reduces the data so that there are two dimensions (see Figure 3-13). You can then apply principal component analysis, which determines how latent variables (also called *components*) best explain the model.

Listing 3-19. Dimension Reduction

```
pca2 = PCA(n_components=2)
pca2.fit(std_df)
x_3d = pca2.transform(std_df)
plt.figure(figsize=(8,6))
```

```
plt.scatter(x_3d[:,0], x_3d[:,1], c=df['gdp_growth'],cmap="viridis",s=350)
plt.xlabel("y")
plt.title("2 dimensional data")
plt.show()
```

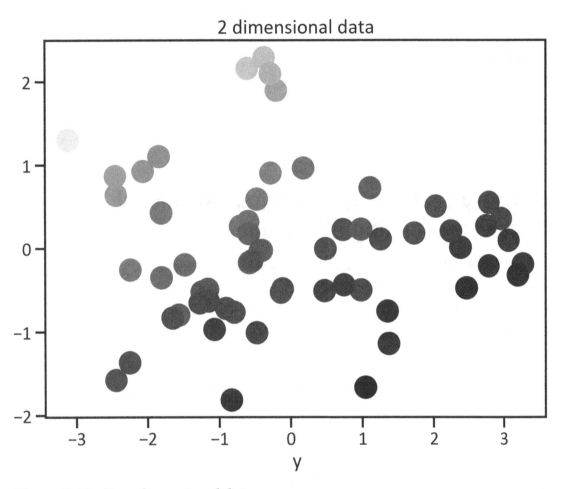

Figure 3-13. *Two-dimensional data*

Figure 3-13 shows the reduced data in a two-dimensional scatter plot. Listing 3-20 generates X and Y arrays and then splits the data into training and test data by applying the 80:20 split ratio. Finally, it standardizes those arrays.

Listing 3-20. Data Preprocessing

```
x = df.iloc[::,0:3]
y = df.iloc[::,-1]
from sklearn.model_selection import train_test_split
from sklearn.preprocessing import StandardScaler
x_train, x_test, y_train, y_test = train_test_split(x,y,test_
size=0.2,random_state=0)
scaler = StandardScaler()
x_train = scaler.fit_tran
sform(x_train)
x_test = scaler.transform(x_test)
```

Ordinary Least-Squares Regression Model Development Using Statsmodels

Listing 3-21 develops a multivariate ordinary least-squares regression model by applying the statsmodels library to determine whether there is a significant difference in the relationship between the U.S. social contributions, the lending interest rate, and the final consumption expenditure. Table 3-8 highlights the values of the U.S. final consumption expenditure that the model predicts.

Listing 3-21. Develop a Multivariate Model Applying Statsmodels

```
import statsmodels.api as sm
x_constant = sm.add_constant(x_train)
x_test = sm.add_constant(x_test)
model = sm.OLS(y_train,x_constant).fit()
pd.DataFrame(model.predict(x_constant), columns = ["Predicted FCE"]).head()
```

Table 3-8. *Predictions (Statsmodels)*

	Predicted FCE
0	1.785433e+12
1	3.810297e+12
2	1.727029e+12
3	7.169544e+12
4	1.272881e+13

Table 3-8 shows the data points that the ordinary least-squares regression model predicts. Listing 3-22 returns Table 3-9. (This profile table provides sufficient information about how the model performs.)

Listing 3-22. Model Evaluation

```
model.summary()
```

Table 3-9. *OLS Model Results*

Dep. Variable	fin_cons	R-squared	0.810
Model	OLS	Adj. R-squared:	0.797
Method	Least Squares	F-statistic:	62.48
Date	Wed, 04 Aug 2021	Prob (F-statistic):	6.70e-16
Time	04:14:27	Log-Likelihood:	-1426.8
No. Observations	48	AIC:	2862.
Df Residuals	44	BIC:	2869.
Df Model	3		
Covariance Type	nonrobust		

(continued)

Table 3-9. (*continued*)

	coef	std err	t	P>ltl	[0.025	0.975]
const	6.953e+12	2.96e+11	23.471	0.000	6.36e+12	7.55e+12
x1	3.858e+12	4.28e+11	9.015	0.000	3e+12	4.72e+12
x2	-1.845e+11	3.91e+11	-0.472	0.639	-9.72e+11	6.03e+11
x3	-1.617e+11	3.35e+11	-0.483	0.631	-8.36e+11	5.13e+11
Omnibus	55.713		**Durbin-Watson**		2.186	
Prob(Omnibus)	0.000		Jarque-Bera (JB)		278.920	
Skew	-3.119		Prob(JB)		2.71e-61	
Kurtosis	13.027		Cond. No.		2.51	

Table 3-9 shows that the ordinary least-squares regression model explains 81.0% of the variability in the model. This means that the model is an exemplary learner. It also shows that only the relationship between the U.S. social contributions and the final consumption expenditure is significant.

It also indicates that:

- For each unit change in the U.S. social contributions, the U.S. final consumption expenditure increases by 3.858e+12.

- For each unit change in the U.S. lending interest rate, the U.S. final consumption expenditure decreases by 1.845e+11.

- For each unit change in the U.S. GDP growth, the U.S. final consumption expenditure decreases by -1.617e+11.

Last, the ordinary least-squares model explains 90% of the variability in the data. This means that the model is exemplary in predicting the U.S. final consumption expenditure.

Residual Analysis

Figure 3-14 shows the distribution of residuals; the code is shown in Listing 3-23.

Listing 3-23. Residual Analysis

```
pd.DataFrame(model.resid,columns=["Residuals"]).plot(kind="box")
plt.title("Model 1 residuals distribution")
plt.ylabel("Values")
plt.show()
```

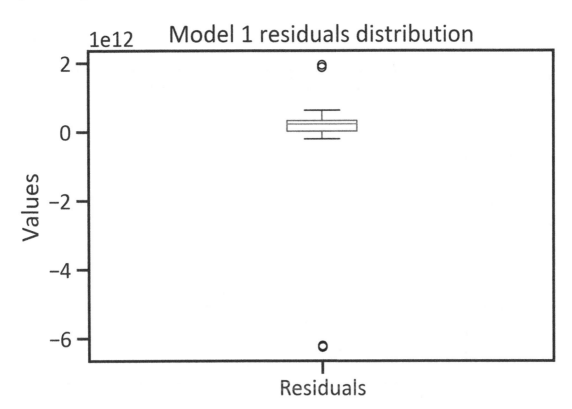

Figure 3-14. *Model residual distribution*

Figure 3-14 indicates that the mean value of the residuals is close to 0, but there are outliers in the residuals.

Residual Autocorrelation

Listing 3-24 determines the series correlation between residuals (see Figure 3-15).

Listing 3-24. Residual Autocorrelation

```
from statsmodels.graphics.tsaplots import plot_acf
plot_acf(model.resid)
plt.title("Model 1 residuals autocorrelation")
plt.ylabel("ACF")
plt.show()
plt.show()
```

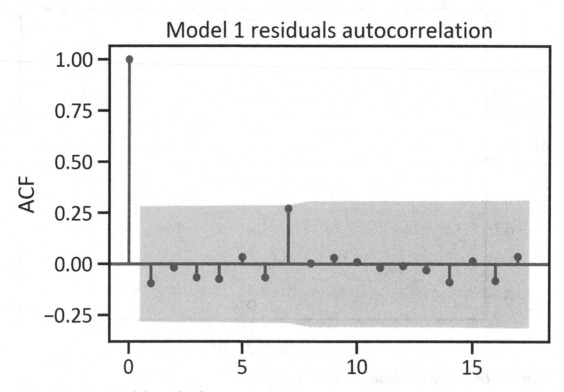

Figure 3-15. *Model residual autocorrelation*

Figure 3-15 indicates that no autocorrelation coefficients exceed the statistical boundary.

Ordinary Least-Squares Regression Model Development Using Scikit-Learn

This section discusses a multivariate ordinary least-squares regression model that uses the scikit-learn library. See Listing 3-25.

Listing 3-25. Data Preprocessing

```
x = df.iloc[::,0:3]
y = df.iloc[::,-1]
from sklearn.model_selection import train_test_split
from sklearn.preprocessing import StandardScaler
x_train, x_test, y_train, y_test = train_test_split(x,y,test_
size=0.2,random_state=0)
scaler = StandardScaler()
x_train = scaler.fit_transform(x_train)
x_test = scaler.transform(x_test)
```

Listing 3-26 trains the multivariate ordinary least-squares regression model.

Listing 3-26. Train the Default Model

```
from sklearn.linear_model import LinearRegression
lm = LinearRegression()
lm.fit(x_train,y_train)
```

Cross-Validation

Listing 3-27 finds the cross-validation score using the cross_val_score() method.

Listing 3-27. Cross-Validation

```
from sklearn.model_selection import train_test_split, cross_val_score
def get_val_score(scores):
    scores = cross_val_score(scores, x_train, y_train, scoring="r2")
    print("CV mean: ", np.mean(scores))
    print("CV std: ", np.std(scores))
```

```
get_val_score(lm)
CV mean:  0.7815577376481077
CV std:  0.1706291983493402
```

These findings show that the mean of the cross-validation scores is 0.7816 and the independent data points of the cross-validation score deviate from the mean by 0.1706.

Hyperparameter Optimization

Listing 3-28 optimizes the hyperparameters using the GridSearchCV() method. See Listings 3-29 and 3-30.

Listing 3-28. Hyperparameter Optimization

```
from sklearn.model_selection import GridSearchCV
param_grid = {"copy_X":[False,True], "fit_intercept":[False,True], "n_
jobs":[-5,-3,3,5], "normalize":[False,True]}
grid_model = GridSearchCV(estimator=lm, param_grid=param_grid)
grid_model.fit(x_train,y_train)
print("Best scores: ", grid_model.best_score_, "Best parameters: ", grid_
model.best_params_)
Best scores:  0.7815577376481078 Best parameters:  {'copy_X': False, 'fit_
intercept': True, 'n_jobs': -5, 'normalize': True}
```

Listing 3-29. Ordinary Least-Squares Regression Model Development

```
lm = LinearRegression(copy_X= False, fit_intercept= True, n_jobs= -5,
normalize=False)
lm.fit(x_train,y_train)
```

Listing 3-30. Estimate the Intercept

```
lm.intercept_
6952809734166.666
```

When you hold the U.S. social contributions, the lending interest rate, and the GDP growth constant, the mean value of the predicted U.S. final consumption expenditure is 6952809734166.667. Listing 3-31 estimates the coefficients.

Listing 3-31. Estimate the Coefficients

```
lm.coef_
array([ 3.85798531e+12, -1.84542533e+11, -1.61721467e+11])
```

This shows that:

- For each unit change in the U.S. social contributions, the U.S. final consumption expenditure increases by 3.85798531e+12.

- For each unit change in the U.S. lending interest rate, the U.S. final consumption expenditure decreases by 1.84542533e+11.

- For each unit change in the U.S. GDP growth, the U.S. final consumption expenditure decreases by 1.61721467e+11.

Residual Analysis

Listing 3-32 compares the actual data points and the predicted data points of the U.S. final consumption expenditures side-by-side (see Table 3-10).

Listing 3-32. The Actual and Predicted U.S. Final Consumption Expenditures

```
actual_values = pd.DataFrame(df["fin_cons"][:13])
actual_values.columns = ["Actual FCE"]
predicted_values  = pd.DataFrame(lm.predict(x_test))
predicted_values.columns = ["Predicted FCE"]
predicted_values.index = actual_values.index
actual_and_predicted_values = pd.concat([actual_values,predicted_
values],axis=1)
actual_and_predicted_values.index = actual_values.index
actual_and_predicted_values
```

Table 3-10. *The Actual and Predicted U.S. Final Consumption Expenditures*

Date	Actual FCE	Predicted FCE
2020-01-01	5.829066e+12	5.596803e+12
2019-01-01	3.403470e+12	3.196027e+12
2018-01-01	7.173358e+12	7.279473e+12
2017-01-01	5.245942e+12	5.214464e+12
2016-01-01	1.227281e+13	1.111828e+13
2015-01-01	1.688457e+13	1.513641e+13
2014-01-01	3.624267e+12	3.595279e+12
2013-01-01	7.173358e+12	7.375272e+12
2012-01-01	2.204317e+12	2.431846e+12
2011-01-01	7.144856e+12	6.626573e+12
2010-01-01	1.543082e+13	1.412346e+13
2009-01-01	1.269598e+13	1.137305e+13
2008-01-01	4.757180e+12	4.747595e+12

Listing 3-33 determines the actual data points and the predicted data points of the U.S. final consumption expenditures in current U.S. dollars (see Figure 3-16).

Listing 3-33. Plot the Actual and Predicted U.S. Final Consumption Expenditures

```
actual_and_predicted_values.plot(lw=4)
plt.xlabel("Date")
plt.title("The U.S. actual and predicted FCE")
plt.ylabel("FCE")
plt.legend(loc="best")
plt.show()
```

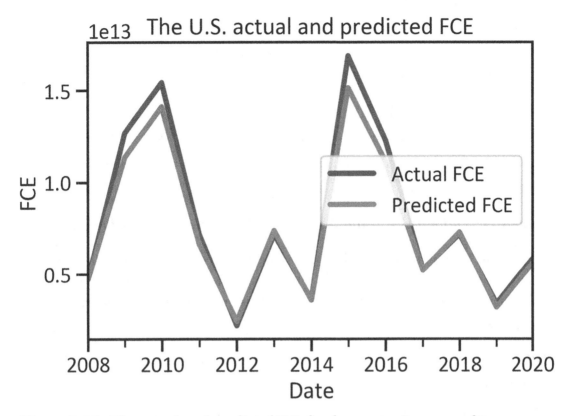

Figure 3-16. *The actual and predicted U.S. final consumption expenditure*

Figure 3-16 plots the actual data points and those that the ordinary least-squares regression model predicts. See Listing 3-34. Figure 3-17 shows the distribution of residuals.

Listing 3-34. Ordinary Least-Squares Regression Model Residual Distribution

```
y_pred = lm.predict(x_test)
residuals = y_test - y_pred
residuals = pd.DataFrame(residuals)
residuals.columns = ["Residuals"]
residuals.plot(kind="box",color="navy")
plt.title("Model 2 residuals distribution")
plt.ylabel("Values")
plt.show()
```

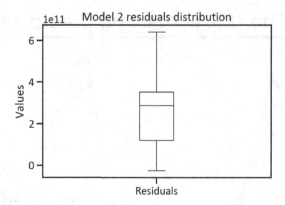

Figure 3-17. *Ordinary least-squares regression model residual distribution*

Figure 3-17 demonstrates that the mean value of the residuals is close to 0. Therefore, you observe autocorrelation in residuals. Autocorrelation refers to the series correlation (see Listing 3-35). Figure 3-18 shows the autocorrelation coefficients of residuals across varying lags.

Listing 3-35. Ordinary Least-Squares Regression Model Residual Autocorrelation

```
from statsmodels.graphics.tsaplots import plot_acf
plot_acf(residuals)
plt.title("Model 2 residuals autocorrelation")
plt.ylabel("ACF")
plt.show()
```

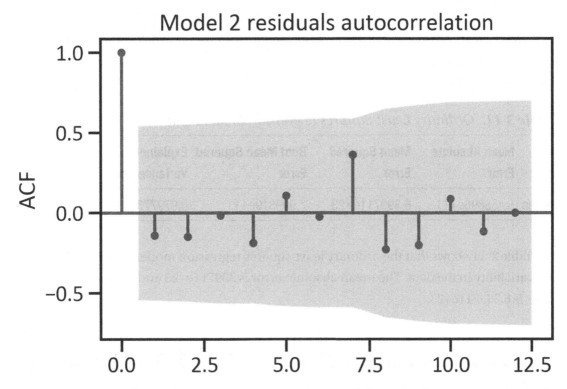

Figure 3-18. *Ordinary least-squares regression model residual autocorrelation*

Figure 3-18 demonstrates that residual autocorrelation coefficients are not outside statistical control. The ordinary least-squares regression model is exemplary at predicting the response variable. See Listing 3-36. Table 3-11 lists other ordinary least-squares regression model performance metrics like the mean absolute error, mean square error, root mean square error, explained variance, and R^2.

Listing 3-36. Key Model Performance Results

```
from sklearn import metrics
MAE = metrics.mean_absolute_error(y_test,y_pred)
MSE = metrics.mean_squared_error(y_test,y_pred)
RMSE = np.sqrt(MSE)
EV = metrics.explained_variance_score(y_test,y_pred)
R2 = metrics.r2_score(y_test,y_pred)
lmmodelevation = [[MAE,MSE,RMSE,EV,R2]]
```

```
lmmodelevationdata = pd.DataFrame(lmmodelevation, index=["Values"],
columns=("Mean absolute error", "Mean squared error", "Root mean squared
error", "Explained variance score", "R-squared"))
lmmodelevationdata
```

Table 3-11. *Ordinary Least-Squares Regression Model Evaluation*

	Mean Absolute Error	Mean Squared Error	Root Mean Squared Error	Explained Variance Score	R^2
Values	5.458896e+11	6.399711e+23	7.999819e+11	0.979773	0.969552

Table 3-11 shows that the ordinary least-squares regression model explains 97% of the variability in the data. The mean absolute error is 399711e+23 and the mean squared error is 6.399711e+23.

Ordinary Least-Squares Regression Model Learning Curve

Figure 3-19 shows how the ordinary least-squares model learns to explain the variability in the relationship among the U.S. social contributions, the lending interest rate, the GDP growth, and the increasing and decreasing final consumption expenditure. See Listing 3-37.

Listing 3-37. Ordinary Least-Squares Regression Model Learning Curve

```
from sklearn.model_selection import learning_curve
trainsizelogreg, trainscorelogreg, testscorelogreg = learning_curve(lm,
x,y,cv=5, n_jobs=-1, scoring="r2", train_sizes=np.linspace(0.1,1.0,50))
trainscorelogreg_mean = np.mean(trainscorelogreg,axis=1)
trainscorelogreg_std = np.std(trainscorelogreg,axis=1)
testscorelogreg_mean = np.mean(testscorelogreg,axis=1)
testscorelogreg_std = np.std(testscorelogreg,axis=1)
fig, ax = plt.subplots()
plt.plot(trainsizelogreg,trainscorelogreg_mean, label="Training R2 score",
color="red",lw=4)
```

```
plt.plot(trainsizelogreg,testscorelogreg_mean, label="Validation R2
score",color="navy",lw=4)
plt.title("OLS learning curve")
plt.xlabel("Training set size")
plt.ylabel("R2 score")
plt.legend(loc=4)
plt.show()
```

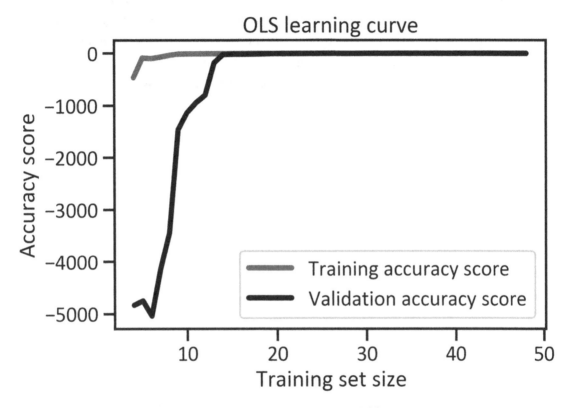

Figure 3-19. *Ordinary least-squares regression model learning curve*

Figure 3-19 shows that the cross-validation R^2 score is constant, and the training R^2 score is low in the first data point, but it increases thereafter.

Conclusion

This chapter used the ordinary least-squares regression model to investigate whether there is a significant difference between the U.S. social contributions (current LCU), the lending interest rate (as a percentage), the GDP growth (as an annual percentage), and the final consumption expenditure (in current U.S. dollars). The model found an extreme positive correlation between the U.S. social contributions and final consumption, a moderate negative correlation between the U.S. lending interest rate and final consumption, and a moderate negative correlation between the U.S. GDP growth and final consumption. The `statsmodels` model and the `scikit-learn` model do not violate regression assumptions and are both exemplary in predicting the U.S. final consumption expenditure, but the `scikit-learn` model has greater predictive power.

CHAPTER 4

Forecasting Growth

Time series analysis involves examining patterns and trends in sequential data to forecast the values of a series. There is a myriad of time series models, including Autoregressive Moving Average (ARIMA) (p, d, q), which applies linear transformation between preceding and current values (autoregressive), integrative (random walk), and moving averages. One ARIMA model includes seasonality, called Seasonal ARIMA (P, D, Q) x (p, d, q). This chapter considers the additive model, which recognizes non-linearity through smooth parameters. For the purposes of this book, you can consider time series analysis an extension of linear regression, as it studies sequential data. The principal difference is that, in time series analysis, you order data over time, which makes assumptions much stricter than in linear regression models. Before training the model, you must test for *stationarity* using the Augmented Dickey-Fuller test, test for the presence of white noise, and test for autocorrelation.

This chapter investigates these trends in the market value of finished goods and services produced in the U.S. annually (see Equation 4-1):

- A guiding movement—Persistent movement uphill is an upward trend and a persistent movement downward is a downward trend

- Seasonality—A persistent year-to-year movement (either uphill or downhill)

- Irregular components—Also called residual components

Lastly, the model forecasts future economic activities. Note that it does not describe the central tendency of the data because the preceding chapter did so.

This chapter uses the Python library called Prophet, which was developed by Facebook. It is grounded on the additive model.

Before you proceed, be sure that you have the `fbprophet` library installed in your environment. To install the `fbprophet` library in a Python environment, use `pip install fbprophet`. Equally, to install the library in a Conda environment, use `conda install -c`

© Tshepo Chris Nokeri 2022
T. C. Nokeri, *Econometrics and Data Science*, https://doi.org/10.1007/978-1-4842-7434-7_4

conda-forge fbprophet. You also need to install the pystan library. To do so in a Python environment, use pip install pystan and in a Conda environment, use conda install -c conda-forge pystan.

Listing 4-1 extracts the U.S. GDP growth data (as an annual percentage) and plots the series (see Figure 4-1).

Listing 4-1. The U.S. GDP Growth (Annual Percentage) Line Plot

```
import wbdata
import matplotlib.pyplot as plt
%matplotlib inline
country = ["USA"]
indicator = {"FI.RES.TOTL.CD":"gdp_growth"}
df = wbdata.get_dataframe(indicator, country=country, convert_date=True)
df.plot(kind="line",color="green",lw=4)
plt.title("The U.S. GDP growth (annual %)")
plt.xlabel("Dates")
plt.ylabel("GDP growth (annual %)")
plt.legend(loc="best")
plt.show()
```

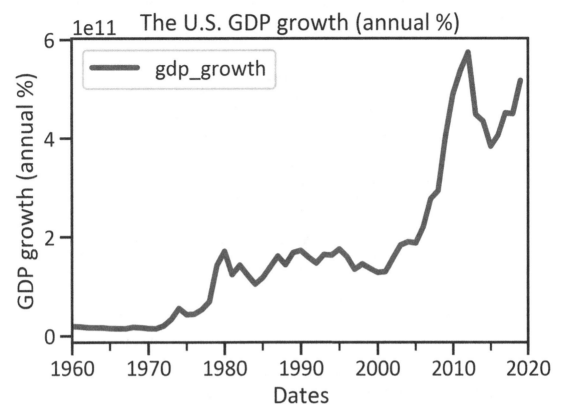

Figure 4-1. *The U.S. GDP growth line plot*

Figure 4-1 demonstrates an upward trend in the GDP growth in the United States since the early 1960s. In early 2012, the U.S. GDP growth declined and then regained momentum in 2015. Listing 4-2 substitutes any missing values with the mean value.

Listing 4-2. Replacing Missing Values with the Mean Value

```
df["gdp_growth"] = df["gdp_growth"].fillna(df["gdp_growth"].mean())
```

Descriptive Statistics

Listing 4-3 retrieves Table 4-1, which outlines the descriptive statistics.

Listing 4-3. Descriptive Statistics

```
df.describe()
```

Table 4-1. *Descriptive Statistics*

	gdp_growth
Count	6.100000e+01
Mean	1.822831e+11
Std	1.646309e+11
Min	1.483107e+10
25%	4.416213e+10
50%	1.441768e+11
75%	1.904648e+11
Max	6.283697e+11

Table 4-1 shows that:

- The mean value of GDP growth in the United States is 6.000000e+01.

- The independent data points of GDP growth in the United States deviate from the mean by 1.748483e+11. See Listing 4-4.

Listing 4-4. The U.S. GDP Growth Distribution

```
df.plot(kind="box",color="green")
plt.title("The U.S. GDP growth (annual %) distribution")
plt.ylabel("Values")
plt.show()
```

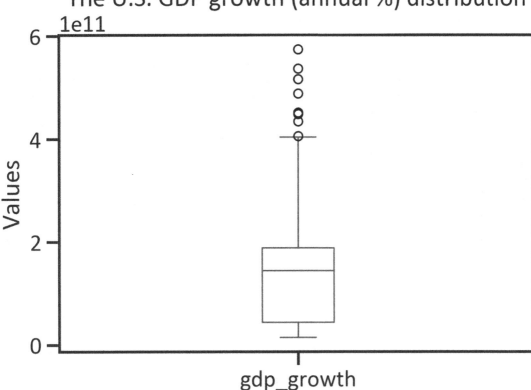

Figure 4-2. *The U.S. GDP growth distribution*

Figure 4-2 shows that there are outliers in the GDP growth data in the United States. Listing 4-5 substitutes the outliers with missing values and checks the new distribution.

Listing 4-5. Replacing the Outliers

```
df['gdp_growth'] = np.where((df["gdp_growth"] > 2.999999e+11),
df["gdp_growth"].mean(),df["gdp_growth"])
df.plot(kind="box",color="green")
plt.title("The U.S. GDP growth (annual %) distribution")
plt.ylabel("Values")
plt.show()
```

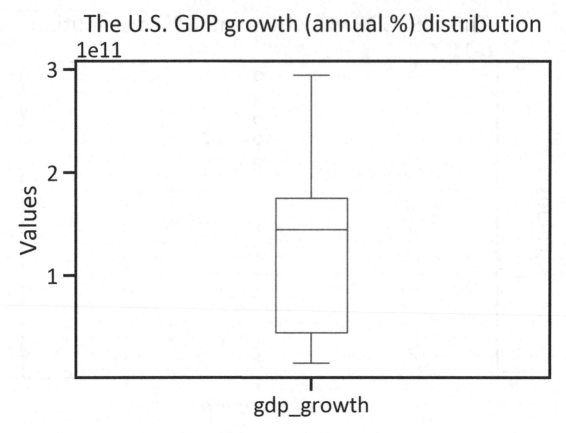

Figure 4-3. *The U.S. GDP growth box plot*

Figure 4-3 shows that there are no more outliers.

Stationarity Detection

Sequential data is stationary if there is randomness in the data (unit root), indicating that the mean of the variable is almost constant or near the mean. Most time series analysis models assume that a series has a unit root. The most common stationarity test is the Augmented Dickey-Fuller test (see Listing 4-6)—it extends the Dickey-Fuller test. When applying the Augmented Dickey-Fuller test, be sure that p-value is significant. Table 4-2 defines the number of lags, the F-statistics %, and the p-value.

Listing 4-6. Augmented Dickey-Fuller Results

```
from statsmodels.tsa.stattools import adfuller
adfullerreport = adfuller(df["gdp_growth"])
```

```
adfullerreportdata = pd.DataFrame(adfullerreport[0:4],
                                  columns = ["Values"],
                                  index=["ADF F% statistics",
                                         "P-value",
                                         "No. of lags used",
                                         "No. of observations"])
adfullerreportdata
```

Table 4-2. *Augmented Dickey-Fuller Results*

Variable	Value
ADF F% Statistics	-0.962163
P-value	0.766813
No. of Lags Used	0.000000
No. of Observations	60.000000

Table 4-2 shows that there is no stationarity in the series, given that the p-value is greater than 0.5. Normally, you would differentiate the series prior to modeling. Note that the test holds a null hypothesis, which suggests that the series is not stationary; it also holds an alternative hypothesis that assumes that the series is stationary.

Random White Noise Detection

Listing 4-7 generates a set of random variables and checks whether the mean value of the data is close to 0 (see Figure 4-4).

Listing 4-7. Random White Noise

```
from pandas.plotting import autocorrelation_plot
randval = np.random.randn(1000)
autocorrelation_plot(randval)
plt.title("Random white noise")
plt.show()
```

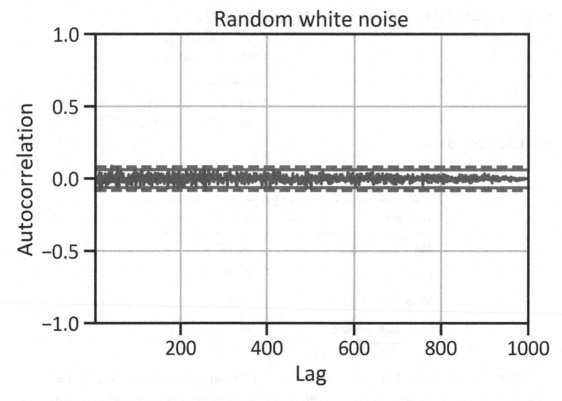

Figure 4-4. *Random white noise*

Figure 4-4 shows that the mean value is around 0. This indicates that the series' patterns are similar across different lags.

Autocorrelation Detection

In simple linear regression, correlation analysis plays a crucial role in model development. Time series analysis, however, is interested in the serial statistical dependence between independent observations (y) and independent observations over time (y_t). Listing 4-8 is an autocorrelation function that makes sense of autocorrelation across varying lags (see Figure 4-5). When autocorrelation coefficients are within the blue region in Figure 4-5, they are significant.

Listing 4-8. The U.S. GDP Growth (Annual %) ACF

```
from statsmodels.graphics.tsaplots import plot_acf
plot_acf(df["gdp_growth"])
plt.title("")
plt.xlabel("Lag")
plt.ylabel("ACF")
plt.title("The U.S. GDP growth (annual %) autocorrelation")
plt.show()
```

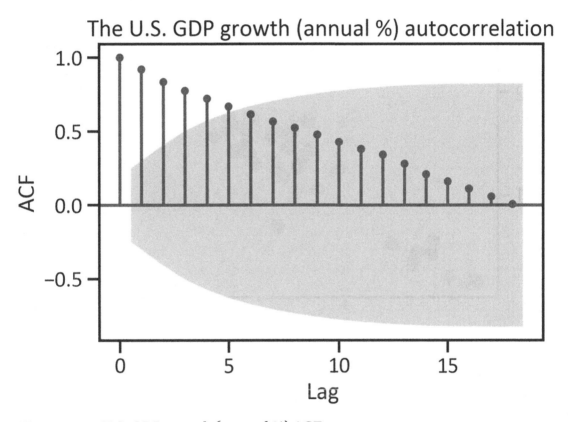

Figure 4-5. *U.S. GDP growth (annual %) ACF*

Listing 4-9 demonstrates independent observations in lag1 (see Figure 4-6). A lag plot helps explain any lags (the space between y_t) against the correlation.

Listing 4-9. The U.S. GDP Growth (Annual %) Lag Plot

```
from pandas.plotting import lag_plot
lag_plot(df["gdp_growth"])
plt.title("The U.S. GDP growth (annual %) lag 1")
plt.show()
```

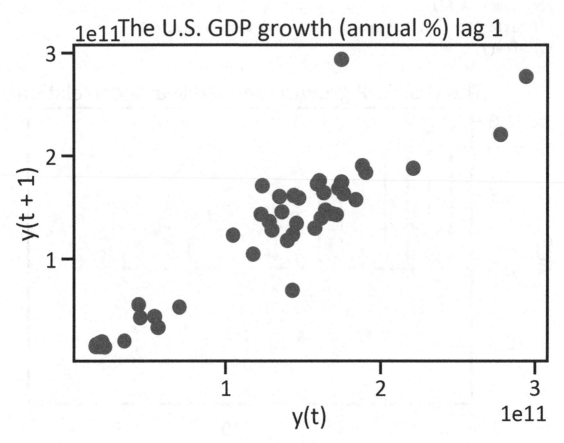

Figure 4-6. *The U.S. GDP growth lag plot*

Figure 4-6 shows that there is a correlational relationship between y_t and $y_{(t+1)}$ in lag1.

Different Univariate Time Series Models

This chapter applies time series analysis to a continuous variable. The following sections contain an overview of the different time series analysis models.

The Autoregressive Integrated Moving Average

Autoregressive Integrated Moving Average is an ARIMA model, which means that it applies linear transformation between preceding and current values (autoregressive), integrative (random walk), and moving averages, which are weights of preceding and current values. It is the most common time series model. This model assumes that there is a linear relationship between the preceding values and that the series is stationary. Not only that, it also considers the difference between current weights and proceeding weighted values.

The Seasonal Autoregressive Integrated Moving Average Model

The Seasonal Autoregressive Integrated Moving Average (SARIMA) extends the ARIMA model, which considers seasonality (a persistent year-to-year movement—either uphill or downhill). Equation 4-1 is the SARIMA formula.

$$Y_t = S_t + T_t + I_t + \varepsilon_i$$
(Equation 4-1)

Where S_t represents seasonality, T_t represents a trend, I_t represents irregularities, and ε_i is the error in terms. SARIMA considers seasonality and does not consider the effects of events like public holidays. To approach this problem, developers frequently apply the additive model.

The Additive Model

The additive model considers the effects of events like public holidays. It adjusts non-linear properties in a function and narrows the forecast to a certain confidence interval, by including smooth parameters and adding non-linearity. The most reliable library for developing an additive model is FB Prophet. Listing 4-10 repurposes the data (see Table 4-3).

Listing 4-10. Repurposing the U.S. GDP Growth (Annual %) Data

```
df = df.reset_index()
df["ds"] = df["gdp_growth"]
df["y"] = df["date"]
df.set_index("date")
```

Table 4-3. *The U.S. GDP Growth (Annual %) Repurposed Data*

Date	gdp_growth	ds	y
2020-01-01	1.822831e+11	1.822831e+11	2020-01-01
2019-01-01	1.822831e+11	1.822831e+11	2019-01-01
2018-01-01	1.822831e+11	1.822831e+11	2018-01-01
2017-01-01	1.822831e+11	1.822831e+11	2017-01-01
2016-01-01	1.822831e+11	1.822831e+11	2016-01-01
...	...	...	...
1964-01-01	1.672549e+10	1.672549e+10	1964-01-01
1963-01-01	1.687865e+10	1.687865e+10	1963-01-01
1962-01-01	1.725246e+10	1.725246e+10	1962-01-01
1961-01-01	1.882553e+10	1.882553e+10	1961-01-01
1960-01-01	1.966416e+10	1.966416e+10	1960-01-01

Additive Model Development

Developing an additive model using the FB Prophet library is fairly simple. Listing 4-11 calls Prophet() and trains the model.

Listing 4-11. Train Additive Model

```
from fbprophet import Prophet
m = Prophet()
m.fit(df)
```

Additive Model Forecast

Listing 4-12 constructs a future dataframe for the next three years and plots data points that the additive model forecasts.

Listing 4-12. Plot U.S. GDP Growth Forecast

```
future = m.make_future_dataframe(periods=3)
forecast = m.predict(future)
m.plot(forecast)
plt.title("The U.S. GDP growth (annual %) forecast")
plt.xlabel("Date")
plt.ylabel("GDP growth (annual %)")
plt.show()
```

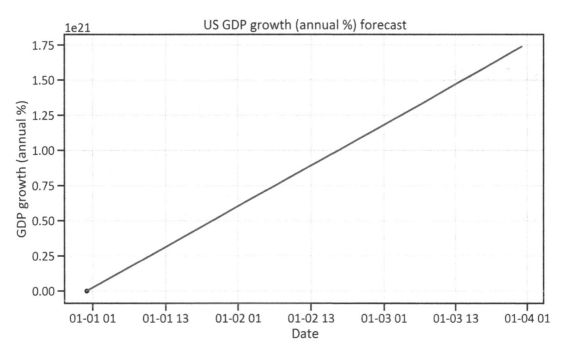

Figure 4-7. *The U.S. GDP growth forecast*

Figure 4-7 indicates an upsurge in U.S. GDP growth. The next section discusses seasonal decomposition—an important aspect of time series analysis. However, given the structure of this data, we do not provide an example. It should be apparent that the figure does not show the dates correctly.

Seasonal Decomposition

Seasonal decomposition involves fragmenting sequential data into a few components, so developers can better comprehend the series. Such components include:

- Seasonality, which represents a permanent pattern over time.

- Trends, which are temporary persistent patterns in a series over time.

- Irregularities, which represent the variability of the U.S. GDP growth data.

This chapter does not cover seasonal decomposition given the frequency at which the data is collected and the fact that there are only a few data points of U.S. GDP growth. They collect data annually, and we cannot break seasonality down into daily and weekly seasonality. It is, however, important to comprehend how seasonal decomposition works.

Conclusion

This chapter introduced a time series analysis model called the additive model. Prior to fitting the model, you learned how to confirm time series analysis assumptions by testing for stationarity, white noise, autocorrelation, and normality. Thereafter, the chapter covered trends, seasonality, and residuals. The model forecasted a dominant upward forthcoming trend in the annual U.S. GDP growth. The next chapter introduces a classification method called *logistic regression*. You use this model when you want to operate predictor variables to generate a binary output value.

CHAPTER 5

Classifying Economic Data Applying Logistic Regression

This chapter applies logistic regression to investigate the nature of the relationship between variables when the response variable is a category (meaning it's limited to only a few classes). This model is the most common non-linear model and it's often called *classification*. It uses a Sigmoid function to model the data.

Figure 5-1 demonstrates that the classifier operates on a predictor variable to generate a binary output value. Equation 5-1 shows the sigmoid equation.

$$s(x) = \frac{1}{1 + e^{-x}} \qquad \text{(Equation 5-1)}$$

You might have recognized that macroeconomic data is often continuous. Worry not, as you can still perform logistic regression on this kind of data. All you need to do is categorize the response variable based on direction (upward or downward). You can base this on a formula of an S-shape, called the *sigmoid shape* (see Figure 5-1).

© Tshepo Chris Nokeri 2022
T. C. Nokeri, *Econometrics and Data Science*, https://doi.org/10.1007/978-1-4842-7434-7_5

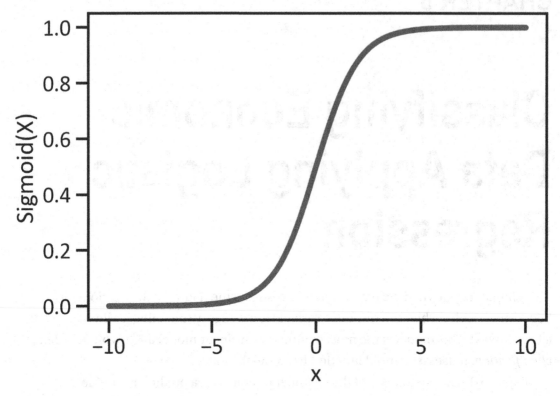

Figure 5-1. *Sigmoid function*

$$( )-1+-x$$

In linear regression, the response variable is continuous. On the other hand, in logistic regression, the response variable is binary and can be applied to multiclass classification. In addition, classifiers do not make extreme assumptions about the linearity and normality of the data.

Context of This Chapter

This chapter predicts the increasing and decreasing life expectancy at birth, in total years, in South Africa by applying a set of other economic indicators—urban population, GNI per capita, Atlas method (in current U.S. dollars), and GDP growth (as an annual percentage). *GNI per capita* is the gross national income on average of a citizen, in this case in South Africa. Table 5-1 outlines South Africa's macroeconomic indicators that this chapter investigates.

Table 5-1. *Macroeconomic Indicators for This Chapter*

Code	Indicator
SP.URB.TOTL	South Africa's urban population
NY.GNP.PCAP.CD	South Africa's GNI per capita, Atlas method (in current U.S. dollars)
NY.GDP.MKTP.KD.ZG	South Africa's GDP growth (as an annual percentage)
SP.DYN.LE00.IN	South Africa's life expectancy at birth, total (in years)

Theoretical Framework

Figure 5-2 provides the framework of this problem.

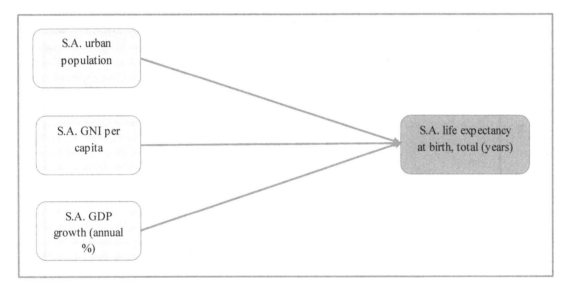

Figure 5-2. *Framework*

Urban Population

The urban population represents the population in areas that have large population densities (i.e., cities). Listing 5-1 and Figure 5-3 show South Africa's urban population.

Listing 5-1. South Africa's Urban Population

```
import wbdata
country  = ["ZAF"]
indicators = {"SP.URB.TOTL":"urban_pop"}
urban_pop = wbdata.get_dataframe(indicators, country=country,
convert_date=True)
urban_pop.plot(kind="line",color="green",lw=4)
plt.title("S.A. urban population")
plt.ylabel("Urban population")
plt.xlabel("Date")
plt.show()
```

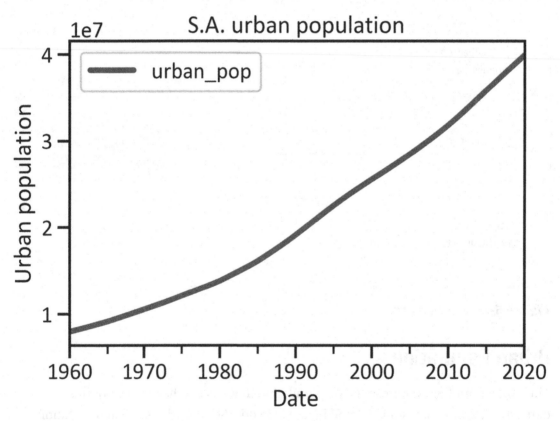

Figure 5-3. *South Africa's urban population*

Figure 5-3 shows a strong upsurge in South Africa's urban population since 1960.

GNI per Capita, Atlas Method

The GNI per capita represents the average income of the citizens of a country. In this example, we apply the Atlas method. Listing 5-2 and Figure 5-4 show South Africa's GNI per capita.

Listing 5-2. South Africa's GNI per Capita

```
country  = ["ZAF"]
indicators = {"NY.GNP.PCAP.CD":"gni_per_capita"}
gni_per_capita = wbdata.get_dataframe(indicators, country=country,
convert_date=True)
gni_per_capita.plot(kind="line",color="orange",lw=4)
plt.title("S.A. GNI per capita")
plt.ylabel("GNI per capita")
plt.xlabel("Date")
plt.show()
```

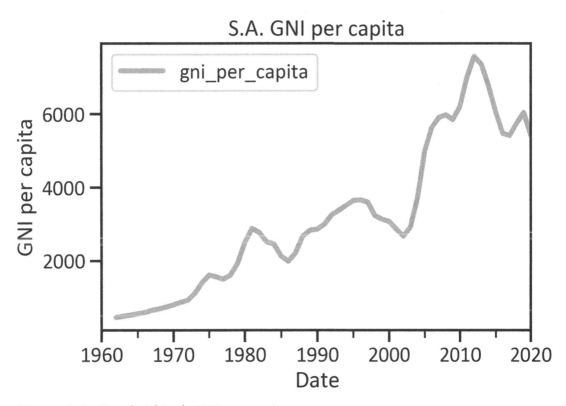

Figure 5-4. *South Africa's GNI per capita*

Figure 5-4 shows a general, extreme upward trend in the average income of South African citizens. It declined in 2012, but then surged upward again in 2015.

GDP Growth

GDP estimates the market value of finished goods and services that households of a country consume annually. It is used to diagnose the health of an economy, as it mirrors a country's economic production. Listing 5-3 and Figure 5-5 show South Africa's GDP growth (as an annual percentage).

Listing 5-3. South Africa's GDP Growth

```
country  = ["ZAF"]
indicators = {"NY.GDP.MKTP.KD.ZG":"gdp_growth"}
gdp_growth = wbdata.get_dataframe(indicators, country=country,
convert_date=True)
gdp_growth.plot(kind="line",color="navy",lw=4)
plt.title("S.A. GDP growth (annual %)")
plt.ylabel("GDP growth (annual %)")
plt.xlabel("Date")
plt.show()
```

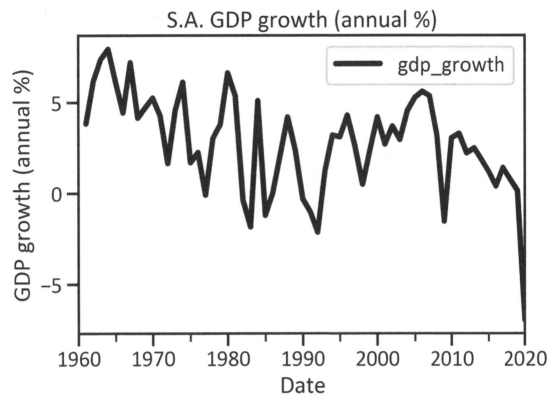

Figure 5-5. *South Africa's GDP growth*

Figure 5-5 shows that the market value of general goods and services that South Africa generates is unstable over time. Between 1960 and 2020, its GDP growth suffered sharp decreases, which leveled off a few years later. The GDP growth was at its peak in the mid-1960s. It reached negative growth in the early 1990s, followed by an upward trend which was corrected in 2008.

Life Expectancy at Birth, Total (in Years)

Life expectancy represents the average number of years a newborn baby is expected to live, provided that the current mortality pattern remains the same. Listing 5-4 and Figure 5-6 show South Africa's life expectancy at birth in years.

Listing 5-4. South Africa's Life Expectancy South Africa's Life Expectancy at Birth Line Plot

```
country  = ["ZAF"]
indicators = {"SP.DYN.LE00.IN":"life_exp"}
life_exp = wbdata.get_dataframe(indicators, country=country, convert_
date=True)
life_exp.plot(kind="line",color="red",lw=4)
plt.title("S.A. life expectancy at birth, total (years)")
plt.ylabel("Life expectancy at birth, total (years)")
plt.xlabel("Date")
plt.show()
```

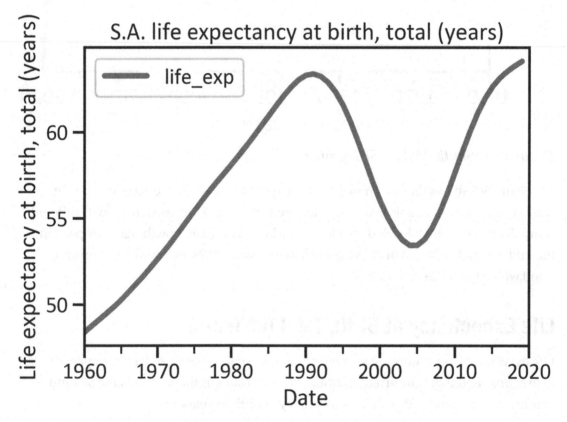

Figure 5-6. *South Africa's life expectancy at birth line plot*

Now that you understand the problem, you can proceed to load the economic indicators of interest into a pandas dataframe. Listing 5-5 retrieves Table 5-2, which lists the first five rows of data.

Listing 5-5. Loading Data

```
country = ["ZAF"]
indicator = {"SP.URB.TOTL":"urban_pop",
            "NY.GNP.PCAP.CD":"gni_per_capita",
            "NY.GDP.MKTP.KD.ZG":"gdp_growth",
            "SP.DYN.LE00.IN":"life_exp"}
df = wbdata.get_dataframe(indicator, country=country, convert_date=True)
df.head()
```

Table 5-2. *South Africa's Macroeconomic Data*

Date	urban_pop	gni_per_capita	gdp_growth	life_exp
2020-01-01	39946775.0	5410.0	-6.959604	NaN
2019-01-01	39149715.0	6040.0	0.152583	64.131
2018-01-01	38348227.0	5750.0	0.787056	63.857
2017-01-01	37540921.0	5410.0	1.414513	63.538
2016-01-01	36726640.0	5470.0	0.399088	63.153

Table 5-2 shows that there are missing values in the data. Listing 5-6 replaces the missing values with the mean value (this is done for each variable in the data).

Listing 5-6. Replace Missing Values with the Mean Value

```
df["urban_pop"] = df["urban_pop"].fillna(df["urban_pop"].mean())
df["gni_per_capita"] = df["gni_per_capita"].fillna(df["gni_per_capita"].mean())
df["gdp_growth"] = df["gdp_growth"].fillna(df["gdp_growth"].mean())
df["life_exp"] = df["life_exp"].fillna(df["life_exp"].mean())
```

Descriptive Statistics

Listing 5-7 detects whether there are outliers in South Africa's urban population data. It also investigates the distribution of the data (see Figure 5-7).

Listing 5-7. South Africa's Urban Population Distribution

```
df["urban_pop"].plot(kind="box",color="green")
plt.title("S.A. urban population distribution")
plt.ylabel("Values")
plt.show()
```

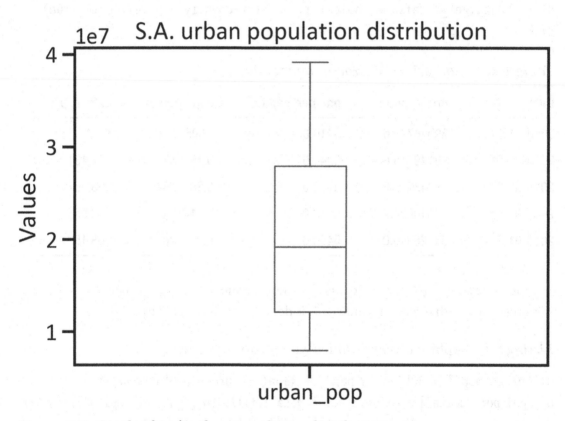

Figure 5-7. *South Africa's urban population distribution*

Figure 5-7 shows that South Africa's urban population data is close to being normally distributed. Listing 5-8 detects whether there are outliers in South Africa's GNI per capita data. It also investigates the distribution of the data (see Figure 5-8).

Listing 5-8. South Africa's GNI Per Capita Distribution

```
df["gni_per_capita"].plot(kind="box",color="orange")
plt.title("S.A. GNI per capita distribution")
plt.ylabel("Values")
plt.show()
```

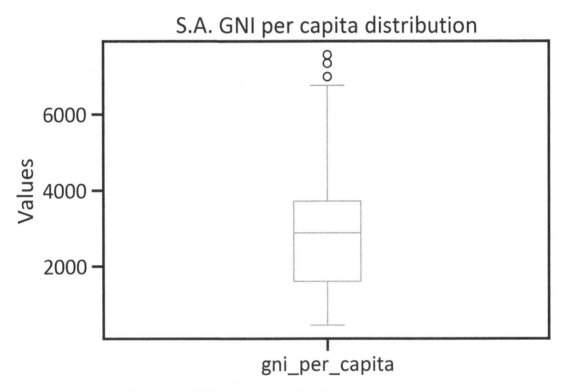

Figure 5-8. *South Africa's GNI per capita distribution*

Figure 5-8 shows that South Africa's GNI per capita data is positively skewed. There are three outliers above the 5 000 value. Listing 5-9 replaces outliers in South Africa's GNI per capita data with the mean value (see Figure 5-9). Besides substituting data using the mean value, you can also apply other techniques, like median value or value near missing values via the k-Nearest Neighbor technique.

Listing 5-9. Replacing Outliers with Mean Values

```
df['gni_per_capita'] = np.where((df["gni_per_capita"] > 5000),df["gni_per_
capita"].mean(),df["gni_per_capita"])
df["gni_per_capita"].plot(kind="box",color="orange")
plt.title("S.A. GNI per capita distribution")
plt.ylabel("Values")
plt.show()
```

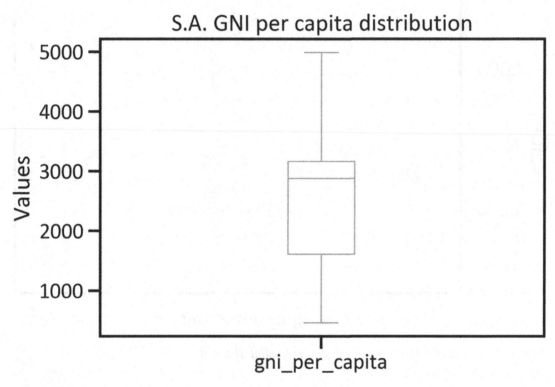

Figure 5-9. *South Africa's GNI per capita distribution*

Listing 5-10 determines whether there are outliers in South Africa's GDP growth data. It also investigates the distribution of the data (see Figure 5-10).

Listing 5-10. South Africa's GDP Growth Distribution

```
df["gdp_growth"].plot(kind="box",color="navy")
plt.title("S.A. GDP growth (annual %) distribution")
plt.ylabel("Values")
plt.show()
```

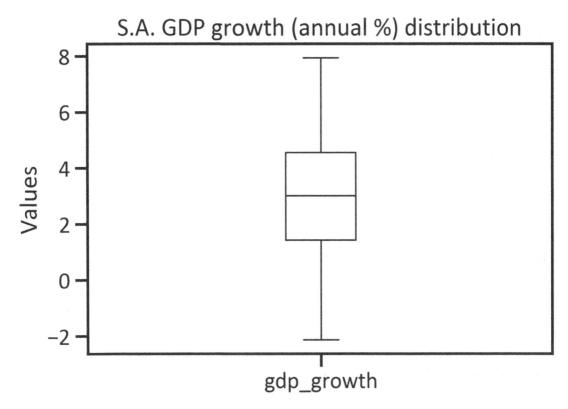

Figure 5-10. *South Africa's GDP growth distribution*

Figure 5-10 shows that South Africa's GDP growth data is normally distributed and there are no outliers. Listing 5-11 detects whether there are outliers in South Africa's life expectancy at birth data. It also shows the distribution of the data (see Figure 5-11).

Listing 5-11. South Africa's Life Expectancy at Birth Distribution

```
df["life_exp"].plot(kind="box",color="red")
plt.title("S.A. life expectancy at birth, total (years) distribution")
plt.ylabel("Values")
plt.show()
```

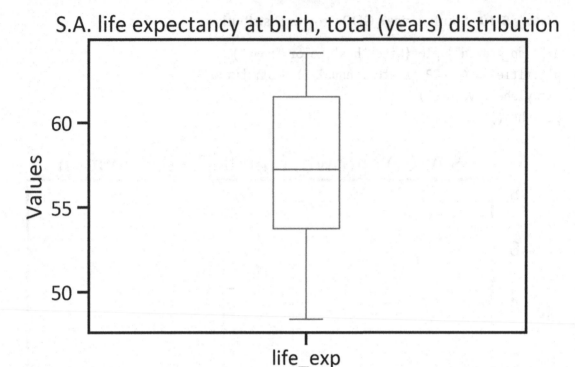

Figure 5-11. *South Africa's Life expectancy at birth distribution*

Figure 5-11 shows that South Africa's life expectancy at birth data has a near-normal distribution, with some slight skewness. Listing 5-12 generates a table that gives us an idea about the central tendency and dispersion of the data (see Table 5-3).

Listing 5-12. Descriptive Summary

```
df.describe().transpose()
```

Table 5-3. *Descriptive Summary*

	Count	Mean	Std	Min	25%	50%	75%	Max
urban_pop	61.0	2.081471e+07	9.698996e+06	7.971773e+06	1.212115e+07	1.914988e+07	2.850619e+07	3.994678e+07
gni_per_ capita	61.0	2.473420e+03	1.074740e+03	4.600000e+02	1.610000e+03	2.880000e+03	3.200508e+03	4.990000e+03
gdp_ growth	61.0	2.775881e+00	2.708735e+00	-6.959604e+00	1.233520e+00	3.014549e+00	4.554553e+00	7.939782e+00
life_exp	61.0	5.724278e+01	4.571705e+01	4.840600e+01	5.374900e+01	5.724278e+01	6.154000e+01	6.413100e+01

Table 5-3 shows that:

- The mean value of South Africa's urban population is 2.081471e+07.

- The GNI per capita, Atlas method is 2.473420e+03 U.S. dollars.

- The GDP growth is 2.775881e+00.

- The life expectancy at birth is 5.724278e+01 years.

- South Africa's urban population deviates from the mean by 9.698996e+06.

- The GNI per capita, Atlas method deviates by 1.074740e+03.

- The GDP growth deviates by 2.708735e+00.

- The life expectancy deviates by 4.571705e+00.

Covariance Analysis

Listing 5-13 and Table 5-4 show how these variables co-vary and vary with themselves.

Listing 5-13. Covariance Matrix

```
dfcov = df.cov()
dfcov
```

Table 5-4. *Covariance Matrix*

	urban_pop	gni_per_capita	gdp_growth	life_exp
urban_pop	9.407052e+13	7.610470e+09	-1.099442e+07	2.381877e+07
gni_per_capita	7.610470e+09	1.155065e+06	-1.089449e+03	2.658211e+03
gdp_growth	-1.099442e+07	-1.089449e+03	7.337248e+00	-6.957400e+00
life_exp	2.381877e+07	2.658211e+03	-6.957400e+00	2.090049e+01

Correlation Analysis

Listing 5-14 finds the correlation coefficients using the Pearson correlation method (see Figure 5-12). This method is useful when the values are continuous. Otherwise, if the values are categorical, you must apply the Kendall or Spearman correlation method.

Listing 5-14. Pearson Correlation Matrix

```
dfcorr = df.corr(method="pearson")
sns.heatmap(dfcorr, annot=True, annot_kws={"size":12},cmap="Blues")
plt.title("Pearson correlation matrix")
plt.show()
```

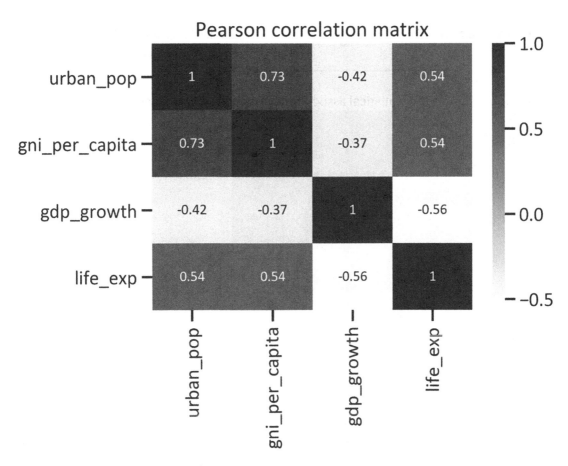

Figure 5-12. *Pearson correlation matrix*

Table 5-5 shows the Pearson correlation coefficients plotted in Figure 5-12.

Table 5-5. *Pearson Correlation Matrix Interpretation*

Variables	Pearson Correlation Coefficient	Findings
South Africa's urban population and life expectancy at birth in years.	0.54	There is a moderate positive correlation between South Africa's urban population and life expectancy at birth.
South Africa's GNI per capita, Atlas method and life expectancy at birth (in years).	0.54	There is a moderate positive correlation between South Africa's GNI per capita, Atlas method and the life expectancy at birth.
South Africa's GDP growth and life expectancy at birth (in years).	-0.56	There is a moderate negative correlation between South Africa's GDP growth and life expectancy at birth in years.

Figure 5-13 plots the statistical association among these variables (see Listing 5-15).

Listing 5-15. Pair Plot

```
sns.pairplot(df)
```

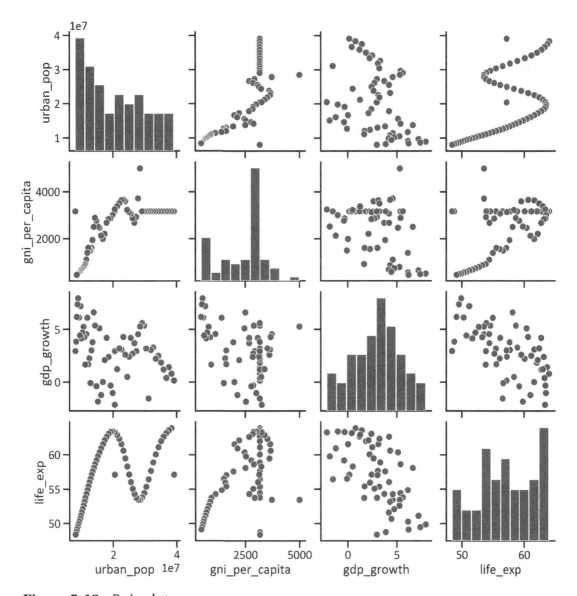

Figure 5-13. *Pair plots*

Figure 5-13 shows histograms of all the variables in the data. The data relating to South Africa's urban population is skewed slightly to the left. That of the GNI per capita, Atlas method, is normal with an exceptional value. The GDP growth data is normal and the life expectancy data is skewed slightly to the right.

115

Correlation Severity Detection

Listing 5-16 and Table 5-6 show the severity of the correlation among variables.

Listing 5-16. Eigen Matrix

```
eigenvalues, eigenvectors = np.linalg.eig(dfcorr)
eigenvalues = pd.DataFrame(eigenvalues)
eigenvalues.columns = ["Eigen values"]
eigenvectors = pd.DataFrame(eigenvectors)
eigenvectors.columns = df.columns
eigenvectors = pd.DataFrame(eigenvectors)
eigenmatrix = pd.concat([eigenvalues, eigenvectors],axis=1)
eigenmatrix.index = df.columns
eigenmatrix
```

Table 5-6. *Eigen Matrix*

	Eigenvalues	urban_pop	gni_per_capita	gdp_growth	life_exp
urban_pop	2.590423	0.526233	0.412356	-0.693983	-0.267261
gni_per_capita	0.738154	0.518674	0.477948	0.705191	-0.072449
gdp_growth	0.266181	-0.440609	0.719198	-0.109331	0.525989
life_exp	0.405242	0.509823	-0.290315	-0.095601	0.804151

Table 5-6 shows that there is no multicollinearity. We use the method to address multicollinearity.

Dimension Reduction

Listing 5-17 and Figure 5-14 show that the variance ratio of the first component is above 2.5. Most variability in the data, and that of the other components, is beneath 1.0.

Listing 5-17. Individual Variance

```
from sklearn.decomposition import PCA
from sklearn.preprocessing import StandardScaler
scaler = StandardScaler()
std_df = scaler.fit_transform(df)
pca = PCA()
pca.fit_transform(std_df)
pca_variance = pca.explained_variance_
plt.figure(figsize=(8, 6))
plt.bar(range(4), pca_variance, align="center", label="Individual
variance")
plt.legend()
plt.ylabel("Variance ratio")
plt.xlabel("Principal components")
plt.title("Individual variance")
plt.show()
```

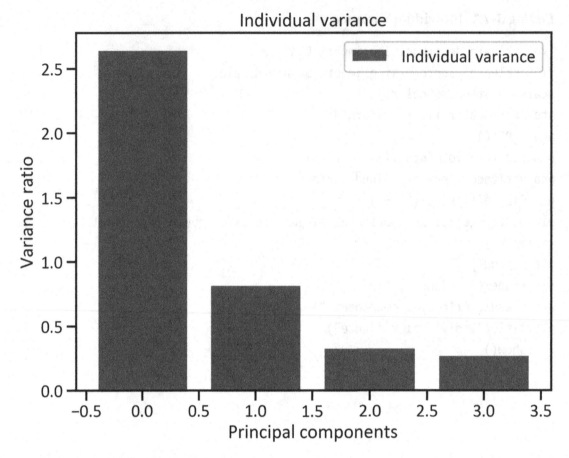

Figure 5-14. *Individual variance*

Listing 5-18 applies principal component analysis and reduces the data in such a way that we can plot it in a two-dimensional scatter plot (see Figure 5-15).

Listing 5-18. Dimension Reduction

```
pca2 = PCA(n_components=2)
pca2.fit(std_df)
x_3d = pca2.transform(std_df)
plt.figure(figsize=(8,6))
plt.scatter(x_3d[:,0], x_3d[:,1], c=df['gdp_growth'],cmap="viridis",s=350)
plt.xlabel("y")
plt.title("2 dimensional data")
plt.show()
```

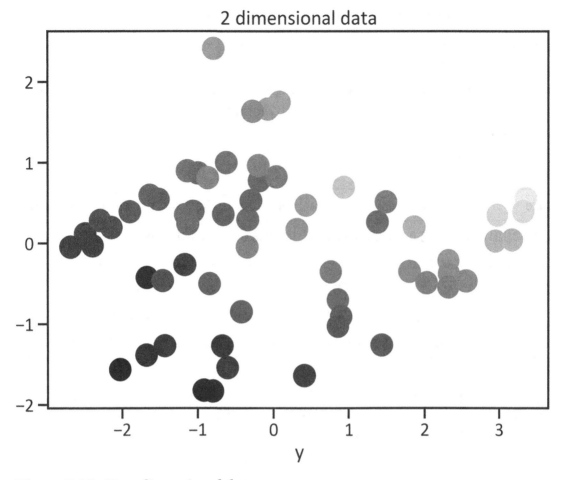

Figure 5-15. *Two-dimensional data*

Figure 5-15 shows how reducing the data scatters it into a two-dimensional scatter plot.

Making a Continuous Variable a Binary

Listing 5-19 retrieves the annual changes in South Africa's life expectancy at birth. It then determines the direction of the series. Last, it *binarizes* the response variable, by applying the get_dummies() method. Binarizing a method means making it categorical (i.e., a "no" or "yes", where you code "no" as 0 and "yes" as 1).

Listing 5-19. Binarize the Response Variable

```
df['pct_change'] = df["life_exp"].pct_change()
df = df.dropna()
df['direction'] = np.sign(df['pct_change']).astype(int)
df["direction"] = pd.get_dummies(df["direction"])
df.head()
```

Table 5-7. *New Dataframe*

Date	urban_pop	gni_per_capita	gdp_growth	life_exp	pct_chage	Direction
2019-01-01	39149715.0	3200.508475	0.152583	64.131	0.120333	0
2018-01-01	38348227.0	3200.508475	0.787056	63.857	-0.004273	1
2017-01-01	37540921.0	3200.508475	1.414513	63.538	-0.004996	1
2016-01-01	36726640.0	3200.508475	0.399088	63.153	-0.006059	1
2015-01-01	35905875.0	3200.508475	1.193733	62.649	-0.007981	1

Table 5-7 includes two new columns, pct_change and Direction. In this chapter, we use Direction as the response variable. We are not going to cover life expectancy and pct_change in this model, so you can delete the columns.

Logistic Regression Model Development Using Scikit-Learn

This section trains the logistic classifier by applying the scikit-learn library. Listing 5-20 first creates an X and Y array, and then splits the data into training and test data. It then standardizes the data.

Listing 5-20. Data Preprocessing

```
from sklearn.preprocessing import StandardScaler
from sklearn.model_selection import train_test_split
x = df.iloc[::,0:3]
scaler = StandardScaler()
```

```
x= scaler.fit_transform(x)
y = df.iloc[::,-1]
x_train, x_test, y_train, y_test = train_test_split(x,y,test_
size=0.2,random_state=0)
```

Listing 5-21 trains the logistic classifier.

Listing 5-21. Train Logistic Classifier by Applying Statsmodels

```
from sklearn.linear_model import LogisticRegression
logregclassifier = LogisticRegression()
logregclassifier.fit(x_train,y_train)
```

Listings 5-22 and 5-23 estimate the intercept and coefficients of the logistic regression model.

Listing 5-22. Intercept

```
logregclassifier.intercept_
array([1.47270205])
```

Listing 5-23. Coefficients

```
logregclassifier.coef_

array([[-0.09342681, -1.02609323,  0.35329756]])
```

Logistic Regression Confusion Matrix

To investigate the logistic regression model's performance, Listing 5-24 creates a *confusion* matrix (see Table 5-8). To evaluate a classification model, you apply evaluation metrics that are different from linear regression models. In classification, you analyze how accurate and precise a model is (see Table 5-10). The initial step involves constructing a confusion matrix (see Table 5-8).

Listing 5-24. Logistic Classifier Confusion Matrix

```
from sklearn import metrics
y_predlogregclassifier = logregclassifier.predict(x_test)
cmatlogregclassifier = pd.DataFrame(metrics.confusion_matrix(y_test,y_
predlogregclassifier), index=["Actual: Decreasing life expectancy",
                                "Actual: Increasing life expectancy"],
                columns = ("Predicted: Decreasing life expectancy",
                                "Predicted: Increasing life expectancy"))
cmatlogregclassifier
```

Table 5-8. *Logistic Classifier Confusion Matrix*

	Predicted: Decreasing Life Expectancy	Predicted: Increasing Life Expectancy
Actual: Decreasing Life Expectancy	0	3
Actual: Increasing Life Expectancy	1	8

Logistic Regression Confusion Matrix Interpretation

Table 5-9 interprets the results of Table 5-8.

Table 5-9. *Logistic Classifier Confusion Matrix Interpretation*

Metric	Interpretation
True Positive	The logistic classifier did not predict a decreasing life expectancy at birth when life expectancy was actually decreasing in South Africa.
True Negative	The logistic classifier predicted an increasing life expectancy at birth when life expectancy was actually increasing in South Africa (three times).
True Positive	The logistic classifier predicted a decreasing life expectancy at birth when life expectancy was actually increasing in South Africa (one time).
False Negative	The logistic classifier predicted an increasing life expectancy at birth when life expectancy was actually decreasing in South Africa (eight times).

Logistic Regression Classification Report

To further investigate the logistic regression model's performance, Listing 5-25 constructs a classification report (see Table 5-10). It outlines how well the model makes accurate and precise estimations.

Listing 5-25. Logistic Regression Classification Report

```
creportlogregclassifier = pd.DataFrame(metrics.classification_report(y_
test,y_predlogregclassifier,output_dict=True))
creportlogregclassifier
```

Table 5-10. *Logistic Regression Classification Report*

	0	1	Accuracy	Macro Avg	Weighted Avg
Precision	0.0	0.727273	0.666667	0.363636	0.545455
Recall	0.0	0.888889	0.666667	0.444444	0.666667
F1-score	0.0	0.800000	0.666667	0.400000	0.600000
Support	3.0	9.000000	0.666667	12.000000	12.000000

Table 5-10 shows that the logistic classifier is accurate 66.67% of the time, and at the most, it is 0% precise when it predicts a decreasing life expectancy. It's 67% precise when it predicts an increasing life expectancy at birth.

Logistic Regression ROC Curve

Listing 5-26 estimates the trade-offs between sensitivity (FPR) and specificity (TPR) as the logistic regression classifier distinguishes between increasing and decreasing life expectancy at birth in South Africa (see Figure 5-16).

Equation 5-2 defines the true positive rate:

$$TPR = \frac{TP}{TP+FN} \qquad \text{(Equation 5-2)}$$

Where TP is the true positive (the inclination of a classifier to make appropriate predictions when they occurred) and the FP is the false positive (the inclination of a classifier to make appropriate predictions when they did not occur).

Equation 5-3 defines the true positive rate:

$$FPR = \frac{FP}{TN+FP} \qquad \text{(Equation 5-3)}$$

Listing 5-26. Logistic Regression ROC Curve

```
y_predlogregclassifier_proba = logregclassifier.predict_proba(x_test)[::,1]
fprlogregclassifier, tprlogregclassifier, _ = metrics.roc_curve(y_test,y_
predlogregclassifier_proba)
auclogregclassifier = metrics.roc_auc_score(y_test,y_predlogregclassifier_
proba)
fig, ax = plt.subplots()
plt.plot(fprlogregclassifier, tprlogregclassifier,label="auc: "+str(auclogr
egclassifier),color="navy",lw=4)
plt.plot([0,1],[0,1],color="red",lw=4)
plt.xlim([0.00,1.01])
plt.ylim([0.00,1.01])
plt.title("Logistic ROC curve")
plt.xlabel("Sensitivity")
plt.ylabel("Specificity")
plt.legend(loc=4)
plt.show()
```

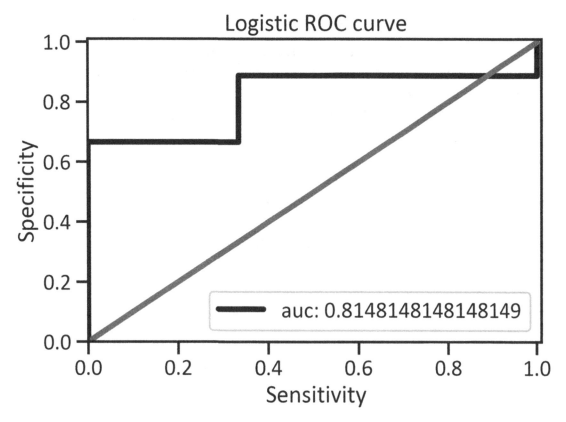

Figure 5-16. *Logistic regression ROC curve*

To detect whether the logistic regression is a suitable class predictor, investigate the area under the curve (AUC) score. If the value exceeds 0.81, the model is considered exemplary. Figure 5-16 shows that on average, the logistic regression classifier is 88.88% accurate when predicting life expectancy in South Africa.

Logistic Regression Precision-Recall Curve

Listing 5-27 estimates the trade-off between precision and the thresholds, as the logistic regression classifier distinguishes between increasing and decreasing life expectancy in South Africa (see Figure 5-17).

Listing 5-27. Logistic Regression Classifier Precision-Recall Curve

```
precisionlogregclassifier, recalllogregclassifier,
thresholdlogregclassifier = metrics.precision_recall_curve(y_test,
y_predlogregclassifier)
apslogregclassifier = metrics.average_precision_score(y_test,
y_predlogregclassifier)
fig, ax = plt.subplots()
plt.plot(precisionlogregclassifier, recalllogregclassifier,label="aps: "
+str(apslogregclassifier),color="navy",lw=4)
plt.axhline(y=0.5, color="red",lw=4)
plt.title("Logistic precision-recall curve")
plt.xlabel("Precision")
plt.ylabel("Recall")
plt.legend(loc=4)
plt.show()
```

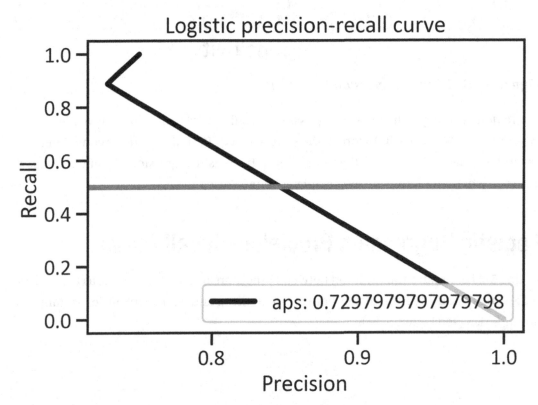

Figure 5-17. *Logistic regression classifier Precision-recall curve*

Figure 5-17 shows that, on average, the logistic regression classifier is 69.44% precise when distinguishing between increasing and decreasing life expectancy in South Africa.

Logistic Regression Learning Curve

Listing 5-28 demonstrates how the logistic regression classifier learns to correctly distinguish between increasing and decreasing life expectancy at birth in South Africa as the training test size increases (see Figure 5-18).

Listing 5-28. Logistic Regression Learning Curve

```
from sklearn.model_selection import learning_curve
trainsizelogregclassifier, trainscorelogregclassifier,
testscorelogregclassifier = learning_curve(logregclassifier, x, y, cv=5,
n_jobs=-1, scoring="accuracy", train_sizes=np.linspace(0.1,1.0,50))
trainscorelogregclassifier_mean = np.mean(trainscorelogregclassifier,axis=1)
trainscorelogregclassifier_std = np.std(trainscorelogregclassifier,axis=1)
testscorelogregclassifier_mean = np.mean(testscorelogregclassifier,axis=1)
testscorelogregclassifier_std = np.std(testscorelogregclassifier,axis=1)
fig, ax = plt.subplots()
plt.plot(trainsizelogregclassifier, trainscorelogregclassifier_mean,
color="red", label="Training accuracy score",lw=4)
plt.plot(trainsizelogregclassifier, testscorelogregclassifier_
mean,color="navy",label="Cross validation accuracy score",lw=4)
plt.title("Logistic classifier learning curve")
plt.xlabel("Training set size")
plt.ylabel("Accuracy")
plt.legend(loc="best")
plt.show()
```

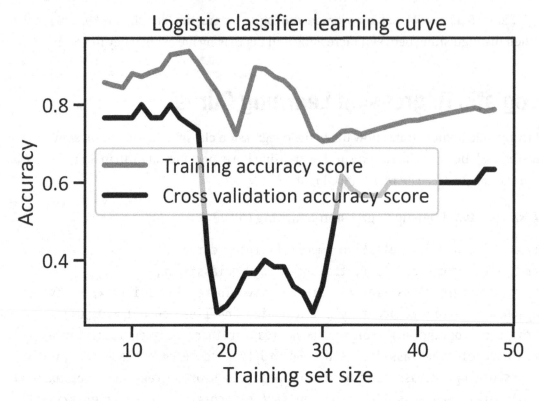

Figure 5-18. *Logistic regression learning curve*

Figure 5-18 shows that, after the first ten data points, the cross-validation accuracy score dips—reaching its lowest peak at the 20th data point—and surges thereafter. In addition, it shows that, after the first ten data points, the training accuracy score is predominantly higher than the cross-validation accuracy score.

Conclusion

This chapter binarized the response variable in such a way that 0 represented a decreasing life expectancy and 1 represented an increasing life expectancy. It used a logistic classifier to operate a set of predictor variables (South Africa's urban population, GNI per capita, Atlas method, and the GDP growth) to predict future classes of life expectancy.

After investigating the findings of this model, we found that it's exemplary at predicting future classes of the response variable (South Africa's life expectancy). Chapter 8 approaches the same problem by applying deep learning models. Chapter 6 covers solving sequential problems in econometrics by applying an unsupervised model.

CHAPTER 6

Finding Hidden Patterns in World Economy and Growth

This chapter introduces decision-making using a Hidden Markov model (HMM). Specifically, it focuses on the Gaussian Mixture model, which is an unsupervised machine learning Hidden Markov model used with time-series data. The beauty of this model is its lack of sensitivity to non-stationary data. After reading this chapter, you will better understand how the Gaussian Mixture model works and will know how to develop one. Note that HMM is an unsupervised learner, which means you can perform an investigation without any predetermined research hypothesis.

This chapter assumes there are two classes, called *states* in Markov decision-making. A series contains continuous values, and at times, you'll want to classify the series. For instance, you can classify a series as having an "upward trend" or a "downward trend," where state 0 represents the upward trend and state 1 represents the downward trend. This is where Markov modeling comes in. Markov modeling involves sequential recognition of current and preceding states, then models to forecast future states. Although it is a useful method for discovering hidden states in a series, there are insufficient metrics for evaluating the models—the most you can get is the mean and covariance at each state.

Before you proceed, make sure you have the `hmmlearn` library installed in your environment. To install the `hmmlearn` library in a Python environment, use `pip install hmmlearn`. Equally, to install the library in a Conda environment, use `conda install -c conda-forge hmmlearn`.

T. C. Nokeri, *Econometrics and Data Science*, https://doi.org/10.1007/978-1-4842-7434-7_6

Applying the Hidden Markov Model

The Hidden Markov model determines preceding hidden states and forecasts future hidden states in a series. For this book, we treat the hidden states as classes. The preceding chapter presented ways of approaching sequential problems using a time series. In addition, it briefly covered binary classification and linear regression. This chapter develops a Gaussian Mixture model to forecast two states (decreasing world GDP growth and increasing world GDP growth). The Gaussian Mixture draws assumptions regarding the data by default. Whenever you come across the word "Gaussian," think of *normality* (variables saturated around the true mean value). This method splits the Gaussian distribution into halves, then determines the probability of a value falling with a certain state. This process happens incrementally.

In summary, this example investigates the hidden patterns in the world GDP growth (as an annual percentage), whether decreasing or increasing. It then applies this model to forecast decreasing and increasing world GDP growth for a specific year. See Listing 6-1 and Table 6-1.

Listing 6-1. Loading World GDP Growth Data

```
import wbdata
country  = ["WLD"]
indicators = {"NY.GDP.MKTP.KD.ZG":"gdp_growth"}
df = wbdata.get_dataframe(indicators, country=country, convert_date=True)
df["gdp_growth"] = df["gdp_growth"].fillna(df["gdp_growth"].mean())
df.head()
```

Table 6-1. *World Economic Growth Data*

Date	gdp_growth
2020-01-01	-3.593456
2019-01-01	2.335558
2018-01-01	3.034061
2017-01-01	3.281329
2016-01-01	2.605572

Descriptive Statistics

Figure 6-1 shows the distribution of the world's GDP growth data using the code in Listing 6-2.

Listing 6-2. World Economic Growth Distribution and Outlier Detection

```
df.plot(kind="box",color="green")
plt.title("World GDP growth (annual %) distribution")
plt.ylabel("Values")
plt.show()
```

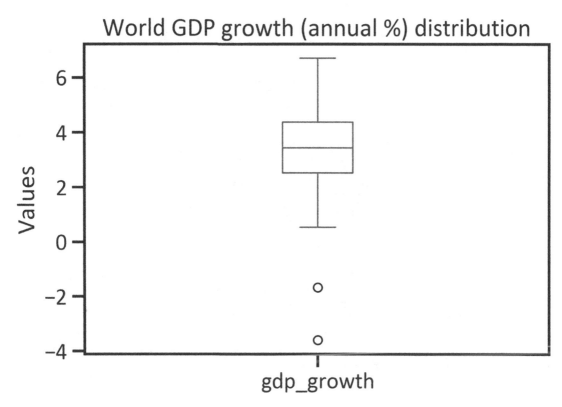

Figure 6-1. *World GDP growth distribution and outlier detection*

Figure 6-1 shows that the distribution of the world's GDP growth data (as an annual percentage) is normally distributed, but there are outliers at the negative axis. Listing 6-3 returns Figure 6-2, which also shows this distribution. See also Listing 6-4.

Listing 6-3. Outlier Removal

```
import numpy as np
df['gdp_growth'] = np.where((df["gdp_growth"] < -1),df["gdp_growth"].
mean(),df["gdp_growth"])
```

Listing 6-4. World GDP Growth Distribution

```
df.plot(kind="hist", color="green")
plt.title("World GDP growth (annual %)")
plt.xlabel("GDP growth (annual %)")
plt.show()
```

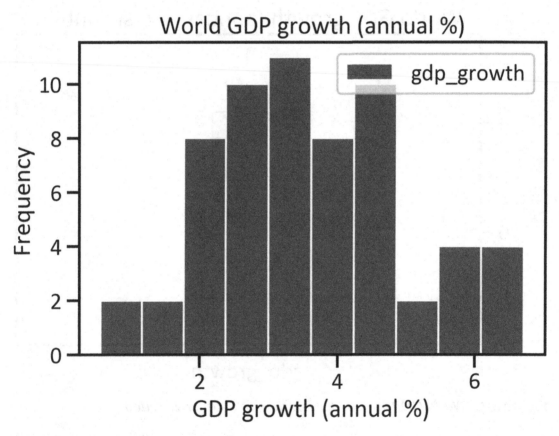

Figure 6-2. *World GDP growth distribution*

Figure 6-2 confirms this normal distribution. Table 6-2 outlines estimates of the central tendency and dispersion using the command in Listing 6-5.

Listing 6-5. World GDP Growth Descriptive Statistics

```
df.describe()
```

Table 6-2. *World GDP Growth Descriptive Statistics*

Variable	gdp_growth
Count	61.000000
Mean	3.566094
Std	1.410267
Min	0.532137
25%	2.605572
50%	3.430696
75%	4.367087
Max	6.710726

Table 6-2 shows that the mean value of the world's GDP growth is 3.566094 and independent observations deviate from the mean value by 1.410267. Likewise, the minimum change in the world's economic production is 0.532137 and the maximum value is 6.710726. Figure 6-3 plots the world's GDP growth from the code in Listing 6-6.

Listing 6-6. World GDP Growth Line Plot

```
df.plot(color="green",lw=4)
plt.title("World GDP growth (annual %) line plot")
plt.xlabel("Date")
plt.ylabel("GDP growth (annual %)")
plt.show()
```

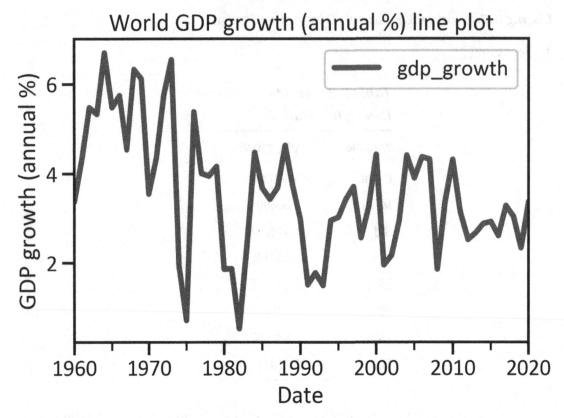

Figure 6-3. *World GDP growth line plot*

Figure 6-3 illustrates the instability in the world GDP growth. We experienced the highest GDP growth in 1963 (at 6.713557) and the lowest in 1982 (at 0.423812). Figure 6-4 shows the log GDP growth using the code in Listing 6-7.

Listing 6-7. World Log GDP Growth Line Plot

```
import pandas as pd
log_df = np.log(df)
log_df = pd.DataFrame(log_df)
log_df.plot(color="green",lw=4)
plt.title("World log GDP growth (annual %) line plot")
plt.xlabel("Date")
plt.ylabel("GDP growth (annual %)")
plt.show()
```

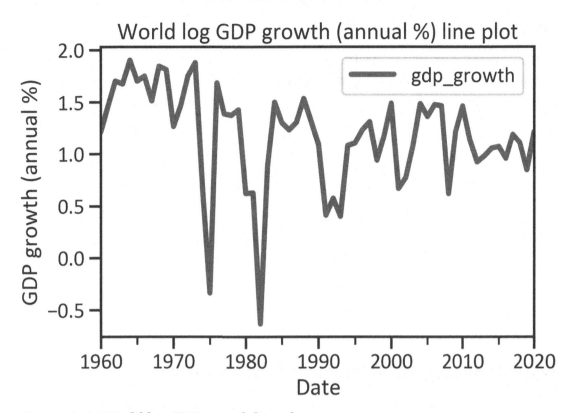

Figure 6-4. *World log GDP growth line plot*

Gaussian Mixture Model Development

Listing 6-8 codes for the Gaussian Mixture model. It assumes that the data comes from a normal distribution.

Listing 6-8. Model Development

```
old_log_df= pd.DataFrame(log_df)
log_df = log_df.values
x = np.column_stack([log_df])
from hmmlearn import hmm
model = hmm.GaussianHMM(n_components=2, tol=0.0001, n_iter=10)
model.fit(x)
```

The code in Listing 6-9 generates the hidden states in the sequence (see Table 6-3).

Listing 6-9. Generate Hidden States

```
hidden_states = pd.DataFrame(model.predict(x), columns = ["hidden_states"])
hidden_states.index = old_log_df.index
hidden_states.head()
```

Table 6-3. *Hidden States*

Date	hidden_states
2020-01-01	1
2019-01-01	1
2018-01-01	1
2017-01-01	1
2016-01-01	1

Representing Hidden States Graphically

To make sense of Table 6-3, refer to Figure 6-5. The code for this process is shown in Listing 6-10.

Listing 6-10. World GDP Growth States

```
increasing_gdp = hidden_states.loc[hidden_states.values == 0]
decreasing_gdp = hidden_states.loc[hidden_states.values == 1]
fig, ax = plt.subplots()
plt.plot(increasing_gdp.index,increasing_gdp.values,".",linestyle="none",
color= "navy",label = "Increasing GDP growth")
plt.plot(decreasing_gdp.index, decreasing_gdp.values,".",linestyle="none",
color = "red",label = "Decreasing GDP growth")
plt.title("World GDP growth (annual %) states")
plt.xlabel("Date")
plt.ylabel("State")
plt.legend(loc="best")
plt.show()
```

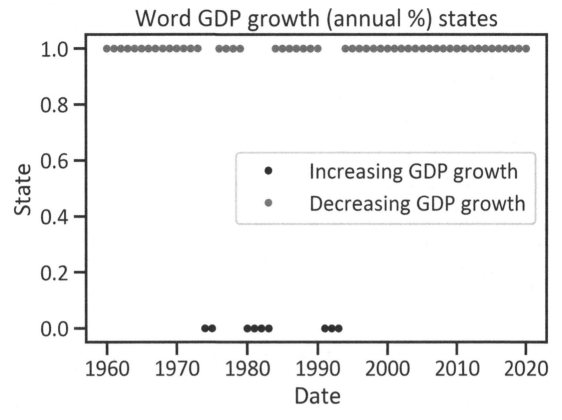

Figure 6-5. *World GDP growth hidden states*

Listing 6-11 returns a scatter plot that demonstrates the dispersion of hidden states (see Figure 6-6).

Listing 6-11. World GDP Growth States Scatter

```
mk_data = old_log_df.join(hidden_states,how = "inner")
mk_data = mk_data[["gdp_growth","hidden_states"]]
up = pd.Series()
down = pd.Series()
mid = pd.Series()
for tuple in mk_data.itertuples():
    if tuple.hidden_states == 0:
        x = pd.Series(tuple.gdp_growth,index = [tuple.Index])
        up = up.append(x)
```

```
    else:
        x = pd.Series(tuple.gdp_growth,index = [tuple.Index])
        down = down.append(x)
        up = up.sort_index()
        down = down.sort_index()
fig, ax = plt.subplots()
plt.plot(up.index, up.values, ".", c = "navy",label = "Increasing GDP
growth")
plt.plot(down.index, down.values,".",c = "red",label = "Decreasing GDP
growth")
plt.title("World GDP growth (annual %) overall")
plt.xlabel("Date")
plt.ylabel("State")
plt.legend(loc="best")
plt.show()
```

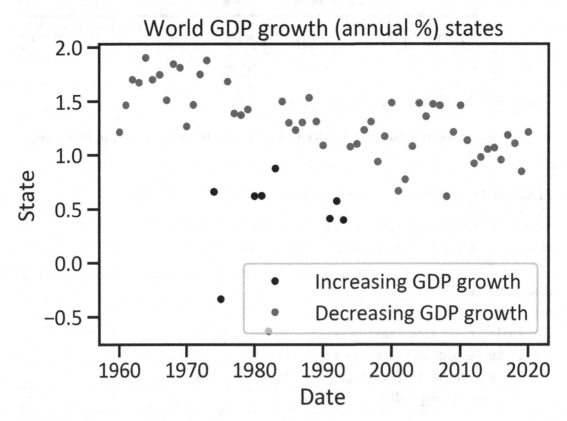

Figure 6-6. *World GDP growth states scatter*

Listing 6-12 counts the hidden states in the sequence and shows them in a pie chart (see Figure 6-7).

Listing 6-12. Hidden States Count

```
binarized_markov = pd.DataFrame(mk_data["hidden_states"])
binarized_markov["hidden_states"]
binarized_markov_data = binarized_markov.replace({0: "Increasing GDP",1:
"Decreasing GDP"})
binarized_markov_data = binarized_markov_data.reset_index()
class_series = binarized_markov_data.groupby("hidden_states").size()
class_series.name = "Count"
class_series.plot.pie(autopct="%2.f",cmap="RdBu")
plt.title("Hidden states pie chart")
plt.show()
```

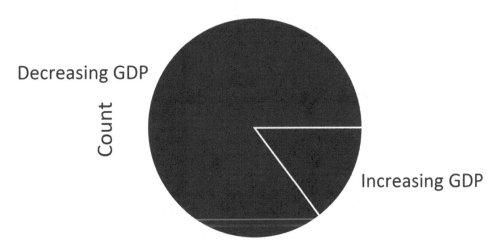

Figure 6-7. *World GDP growth hidden states pie chart*

Figure 6-7 shows that 15% of the forecasted values are increasing GDP growth and 85% are decreasing. Listing 6-13 plots the hidden states across randomly generated samples (Figure 6-8).

Listing 6-13. Hidden States Across Samples

```
num_sample = 1000
sample, _ = model.sample(num_sample)
plt.plot(np.arange(num_sample), sample[:,0],color="green")
plt.title("Hidden states across samples")
plt.xlabel("Samples")
plt.ylabel("States")
plt.show()
```

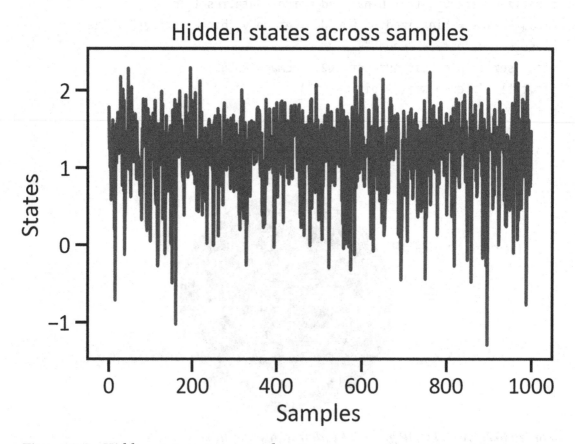

Figure 6-8. *Hidden states across samples*

To make sense of these results, you need to take a closer look at the components.
Listing 6-14 investigates the mean and variance in each order.

Order Hidden States

Listing 6-14. Markov Components

```
for i in range(model.n_components):
    print("{0} order hidden state".format(i))
    print"mean =", model.means_[i])
    print"var =", np.diag(model.covars_[i]))
    print()
```

```
0 order hidden state
0 order hidden state
mean =  [0.58952373]
var =  [0.30547955]

1 order hidden state
mean =  [1.33696442]
var =  [0.0882703]
```

In the 0 order hidden state, the mean is 0.58952373 and the variance is 0.30547955. In the 1 order hidden state, the mean is 1.33696442 and variance is 0.0882703.

Conclusion

This chapter introduced the Gaussian Mixture model, which is a type of Hidden Markov model (HMM). It explored the model using two components, given that there was a predetermined number of states (where 0 represented the first state and 1 represented the second state). It then applied the model to forecast future states in the series data of the world GDP growth (as an annual percentage).

CHAPTER 7

Clustering GNI Per Capita on a Continental Level

This chapter covers unsupervised machine learning. In unsupervised machine learning, you do not fragment data, rather you present the model with the entire data set and allow the model to discover patterns in the data on its own, without any supervision. Given that this chapter covers cluster analysis, it introduces the simplest unsupervised machine learning model, called *k-means.* The k-means method is easy to comprehend. When you're conducting cluster analysis, there is no real response variable—you treat all the variables the same. There is no predictor variable either, so there is no case of prediction/forecasting. You only analyze the patterns in the data.

The k-means model groups values based on their similarities. First, the model finds values that are close to each other and estimates the distance between them. Thereafter, it determines the center of the group of values and stretches those values in such a way that the values are within a cluster. It finds similarities in groups by implementing distance estimation. The most common distance estimation method is Euclidean distance estimation. Equation 7-1 exhibits the Euclidean equation.

$$J = \sum_{i=1}^{k} \sum_{i=1}^{c} \left\| x_i^{J} - C_j \right\|^2 \qquad \text{(Equation 7-1)}$$

Where J represents the objective function, k represents the number of clusters, n represents the number of cases, x_i^{J} represents $Case_j$, and C_j represents the centroids of j.

This model varies from other models in the sense that it assumes that you find the number of clusters based on the analysis of data. To determine the number of clusters, you apply the elbow curve or the scree plot.

T. C. Nokeri, *Econometrics and Data Science*, https://doi.org/10.1007/978-1-4842-7434-7_7

Context of This Chapter

This chapter approaches the problem with the assumption that there are three clusters in the data. It groups African countries by applying these three labels:

- Low GNI (gross national income) per capita

- Moderate GNI per capita

- High GNI per capita

It applies the elbow curve to determine whether this assumption regarding the data is correct. It then uses the k-means model on the entire GNI per capita data of African countries. Given that there are many variables, it does not describe all the data.

To conduct the cluster analysis, you first need to perform descriptive analysis, and then you perform dimension reduction to reduce any high-dimensional data into a few dimensions. Previous chapters covered the dimension reduction technique called Principal Component Analysis (PCA). This technique reduces data so you can represent it in two dimensions by applying a scatter plot. Listing 7-1 retrieves the GNI per capita data for this chapter (see Table 7-1).

Listing 7-1. Load Africa's GNI Per Capita Data

```
import wbdata
countries = ["AGO", "BDI", "BEN", "BWA", "CAF", "CIV",
             "COD", "DZA", "EGY", "ETH", "GAB", "GHA",
              "GIN", "GMB", "GNQ", "KEN", "LBR", "LBY",
             "LSO", "MAR", "MLI", "MOZ", "MWI", "NAM",
             "NER", "NGA", "RWA", "SDN", "SEN","SOM",
             "SWZ", "TCD", "TUN", "TZA", "UGA", "ZAF",
             "ZMB", "ZWE"]
indicators = {"NY.GNP.PCAP.CD":"gni_per_capita"}
df = wbdata.get_dataframe(indicators, country=countries, convert_
date=False)
df.head()
```

Table 7-1. *Africa's GNI Per Capita Data*

Country	Date	gni_per_capita
Angola	2020	2230.0
	2019	2970.0
	2018	3210.0
	2017	3450.0
	2016	3770.0

Listing 7-2 unstacks the data—it changes the index from a multiple index to a single one.

Listing 7-2. Unstack Data

```
df = df["gni_per_capita"]
df = df.unstack(level=0)
```

Listing 7-3 substitutes the missing values with the mean value. Note that this example does not handle outliers. Given that there are so many variables, it would be exhausting to do so in this simple example. In the real world, it is critical to handle outliers, as they can adversely affect conclusions.

Listing 7-3. Replace Outliers with Mean Values

```
df["Algeria"] = df["Algeria"].fillna(df["Algeria"].mean())
df["Angola"] = df["Angola"].fillna(df["Angola"].mean())
df["Benin"] = df["Benin"].fillna(df["Benin"].mean())
df["Botswana"] = df["Botswana"] .fillna(df["Botswana"] .mean())
df["Burundi"] = df["Burundi"].fillna(df["Burundi"] .mean())
df["Central African Republic"] = df["Central African Republic"].
fillna(df["Central African Republic"].mean())
df["Chad"] = df["Chad"].fillna(df["Chad"].mean())
df["Congo, Dem. Rep."] = df["Congo, Dem. Rep."].fillna(df["Congo,
Dem. Rep."].mean())
df["Cote d'Ivoire"] = df["Cote d'Ivoire"].fillna(df["Cote d'Ivoire"].
mean())
```

```
df["Egypt, Arab Rep."] = df["Egypt, Arab Rep."].fillna(df["Egypt, Arab
Rep."].mean())
df["Equatorial Guinea"] = df["Equatorial Guinea"].fillna(df["Equatorial
Guinea"].mean())
df["Eswatini"] = df["Eswatini"].fillna(df["Eswatini"].mean())
df["Ethiopia"] = df["Ethiopia"].fillna(df["Ethiopia"].mean())
df["Gabon"] = df["Gabon"].fillna(df["Gabon"].mean())
df["Gambia, The"] = df["Gambia, The"].fillna(df["Gambia, The"].mean())
df["Ghana"] = df["Ghana"].fillna(df["Ghana"].mean())
df["Guinea"] = df["Guinea"].fillna(df["Guinea"].mean())
df["Kenya"] = df["Kenya"].fillna(df["Kenya"].mean())
df["Lesotho"] = df["Lesotho"].fillna(df["Lesotho"].mean())
df["Liberia"] = df["Liberia"].fillna(df["Liberia"].mean())
df["Libya"] = df["Libya"].fillna(df["Libya"].mean())
df["Malawi"] = df["Malawi"].fillna(df["Malawi"].mean())
df["Mali"] = df["Mali"].fillna(df["Mali"].mean())
df["Morocco"] = df["Morocco"].fillna(df["Morocco"].mean())
df["Mozambique"] = df["Mozambique"] .fillna(df["Mozambique"].mean())
df["Namibia"] = df["Namibia"].fillna(df["Namibia"].mean())
df["Niger"] = df["Niger"].fillna(df["Niger"].mean())
df["Nigeria"] = df["Nigeria"].fillna(df["Nigeria"].mean())
df["Rwanda"] = df["Rwanda"].fillna(df["Rwanda"].mean())
df["Senegal"] = df["Senegal"].fillna(df["Senegal"].mean())
df["Somalia"] = df["Somalia"].fillna(df["Somalia"].mean())
df["South Africa"] = df["South Africa"].fillna(df["South Africa"].mean())
df["Sudan"] = df["Sudan"].fillna(df["Sudan"].mean())
df["Tanzania"] = df["Tanzania"].fillna(df["Tanzania"].mean())
df["Tunisia"] = df["Tunisia"].fillna(df["Tunisia"].mean())
df["Uganda"] = df["Uganda"].fillna(df["Uganda"].mean())
df["Zambia"] = df["Zambia"].fillna(df["Zambia"].mean())
df["Zimbabwe"] = df["Zimbabwe"].fillna(df["Zimbabwe"].mean())
```

Descriptive Statistics

The command in Listing 7-4 retrieves Table 7-2, which demonstrates the central tendency and dispersion of the data.

Listing 7-4. Descriptive Summary

```
df.describe().transpose()
```

Table 7-2. *Descriptive Statistics*

Country	Count	Mean	Std	Min	25%	50%	75%	Max
Algeria	61.0	2180.169492	1482.758119	190.0	1100.000000	1980.000000	2810.000000	5510.0
Angola	61.0	1848.857143	1141.479700	320.0	860.000000	1848.857143	1848.857143	5010.0
Benin	61.0	512.542373	383.386633	90.0	230.000000	370.000000	820.000000	1280.0
Botswana	61.0	2896.271186	2466.075086	80.0	580.000000	2896.271186	4930.000000	7660.0
Burundi	61.0	170.338983	71.483474	50.0	120.000000	170.000000	230.000000	280.0
Central African Republic	61.0	312.542373	138.631324	80.0	240.000000	312.542373	420.000000	570.0
Chad	61.0	352.203390	263.527783	110.0	190.000000	220.000000	470.000000	980.0
Congo, Dem. Rep.	61.0	287.037037	96.191039	110.0	287.037037	287.037037	287.037037	550.0
Cote d'Ivoire	61.0	856.779661	504.435132	160.0	600.000000	760.000000	1070.000000	2290.0
Egypt, Arab Rep.	61.0	1236.666667	919.878615	170.0	550.000000	1130.000000	1420.000000	3440.0
Equatorial Guinea	61.0	4340.256410	4072.093027	130.0	700.000000	4340.256410	4340.256410	14250.0
Eswatini	61.0	2491.470588	884.784592	300.0	1930.000000	2491.470588	3080.000000	4420.0
Ethiopia	61.0	315.526316	174.754088	110.0	210.000000	315.526316	315.526316	890.0
Gabon	61.0	4226.610169	2482.096298	330.0	3100.000000	4226.610169	5520.000000	9330.0
Gambia, The	61.0	477.924528	215.919418	110.0	300.000000	477.924528	660.000000	890.0
Ghana	61.0	652.203390	599.905736	190.0	300.000000	390.000000	652.203390	2230.0
Guinea	61.0	562.121212	137.837556	340.0	490.000000	562.121212	562.121212	1020.0
Kenya	61.0	518.983051	427.316022	100.0	260.000000	390.000000	518.983051	1760.0
Lesotho	61.0	678.909091	406.061049	80.0	410.000000	640.000000	1020.000000	1500.0

Liberia	61.0	458.947368	88.879970	180.0	458.947368	458.947368	458.947368	630.0
Libya	61.0	7765.263158	1585.829298	4550.0	7765.263158	7765.263158	7765.263158	12380.0
Malawi	61.0	214.067797	133.542985	50.0	130.000000	170.000000	300.000000	580.0
Mali	61.0	373.653846	229.876988	60.0	220.000000	300.000000	460.000000	870.0
Morocco	61.0	1517.169811	900.271028	220.0	790.000000	1400.000000	2070.000000	3200.0
Mozambique	61.0	416.785714	98.675454	190.0	416.785714	416.785714	416.785714	690.0
Namibia	61.0	3194.358974	1155.231283	1320.0	2270.000000	3194.358974	3530.000000	5950.0
Niger	61.0	318.135593	136.802468	150.0	220.000000	280.000000	390.000000	600.0
Nigeria	61.0	934.237288	801.430395	100.0	360.000000	630.000000	1350.000000	2940.0
Rwanda	61.0	314.576271	228.752574	40.0	150.000000	270.000000	350.000000	830.0
Senegal	61.0	851.509434	312.396096	310.0	640.000000	851.509434	990.000000	1430.0
Somalia	61.0	147.428571	58.090242	70.0	110.000000	147.428571	147.428571	320.0
South Africa	61.0	3200.508475	1997.563869	460.0	1610.000000	2880.000000	4990.000000	7570.0
Sudan	61.0	578.305085	407.770984	140.0	330.000000	450.000000	720.000000	1540.0
Tanzania	61.0	559.354839	226.501917	160.0	500.000000	559.354839	559.354839	1100.0
Tunisia	61.0	2032.962963	1196.525010	220.0	1160.000000	2032.962963	3100.000000	4160.0
Uganda	61.0	439.459459	187.988882	180.0	290.000000	439.459459	439.459459	850.0
Zambia	61.0	650.000000	445.985799	190.0	360.000000	450.000000	690.000000	1800.0
Zimbabwe	61.0	694.406780	306.688067	260.0	450.000000	670.000000	850.000000	1410.0

Listing 7-5 determines the GNI per capita of the first five countries over time (see Figure 7-1).

Listing 7-5. Africa's GNI Per Capita Line Plot

```
df.iloc[::,0:5].plot()
plt.title("GNI per capita, Atlas method (current US$) line plot")
plt.xlabel("Date")
plt.ylabel("GNI per capita, Atlas method (current US$)")
plt.show()
```

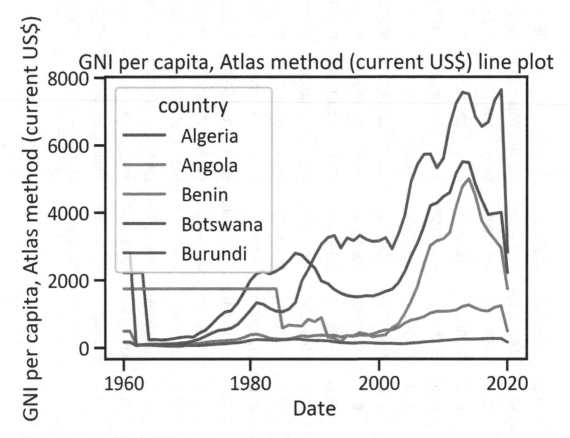

Figure 7-1. *Africa's GNI per capita line plot*

Figure 7-1 shows that, from 1960 to 2020, Burundi has the lowest GNI per capita among the five countries. It also shows that the GNI per capita of Algeria, Angola, Benin, and Botswana surged upward in the early 2000s.

Dimension Reduction

Listing 7-6 applies principal component analysis to reduce the data into two dimensions and then Listing 7-7 graphically represents the reduced data using a two-dimensional scatter plot (see Figure 7-2). Before you perform dimension analysis, you have to *standardize* the data (center the data so the mean value is 0 and the standard deviation is 1).

Listing 7-6. Standardize the Data

```
from sklearn.preprocessing import StandardScaler
scaler = StandardScaler()
std_df = scaler.fit_transform(df)
```

Listing 7-7. Reduce the Data

```
from sklearn.decomposition import PCA
pca2 = PCA(n_components=3)
pca2.fit(std_df)
x_3d = pca2.transform(std_df)
plt.scatter(x_3d[:,0], x_3d[:,2], c=df['South
Africa'],cmap="viridis",s=200)
plt.title("Africa Inequality 2-D Data")
plt.xlabel("y")
plt.show()
```

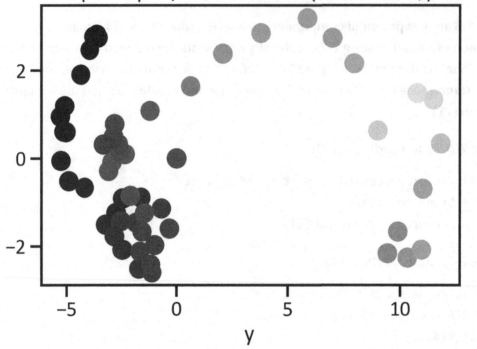

Figure 7-2. *Two-dimensional data*

Figure 7-2 shows the GNI per capita of African countries in a two-dimensional scatter plot.

Cluster Number Detection

Listing 7-8 uses the Elbow curve to identify the number of clusters to include prior to training the k-means model. An elbow plots the *eigenvalue* (a scalar vector after a linear transformation) on the y-axis and the number of *k* on the x-axis.

Listing 7-8. Elbow Curve

```
pca = PCA(n_components=3).fit(std_df)
pca_df = pca.transform(std_df)
pca_df = pca.transform(std_df)
from sklearn.cluster import KMeans
Nc = range(1,20)
```

```
kmeans = [KMeans(n_clusters=i) for i in Nc]
scores = [kmeans[i].fit(pca_df).score(pca_df) for i in range(len(kmeans))]
fig, ax = plt.subplots()
plt.plot(Nc, scores,color="navy",lw=4)
plt.xlabel("No. of clusters")
plt.title("Elbow curve")
plt.ylabel("Score")
plt.show()
```

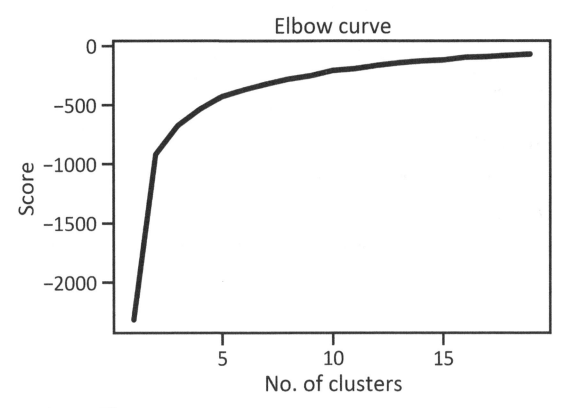

Figure 7-3. *Elbow curve*

Figure 7-3 shows the elbow curve. The x-axis is the number of k (representing the number of clusters), and the y-axis is the distortion (representing the eigenvalues—a scalar retrieved after a linear transformation). The y-axis reveals a lot regarding the severity of correlation among variables in this data, also called WSS (within clusters of sums of squares). To determine how many clusters you must include, you need to determine the *cut-off point*. To select the cut-off point, you need to identify the point at which the elbow curve starts to drastically bend. In this case, the number of clusters is three.

K-Means Model Development

Listing 7-9 trains the k-means model by applying all the data and the cluster number (three).

Listing 7-9. K-Means Model Development

```
kmeans = KMeans(n_clusters=3,
                copy_x=False,
                max_iter= 1,
                n_init= 10,
                tol= 1.0)
kmeans_output = kmeans.fit(pca_df)
kmeans_output
```

Predictions

Listing 7-10 predicts the labels (see Table 7-3)

Listing 7-10. Predictions

```
y_predkmeans = pd.DataFrame(kmeans_output.labels_, columns = ["Clusters"])
y_predkmeans
```

Table 7-3. *Predicted Labels*

	Clusters
0	0
1	0
2	2
3	2
4	2
...	...
56	1
57	1
58	1
59	1
60	1

Listing 7-10 does not reveal much about how the k-means model estimates labels.

Cluster Centers Detection

Listing 7-11 retrieves Table 7-4, which shows the mean value of each cluster.

Listing 7-11. Find Cluster Centers

```
centers = kmeans_output.cluster_centers_
centroids = pd.DataFrame(centers).transpose()
centroids.columns = ["Cluster 1","Cluster 2", "Cluster 3"]
centroids
```

Table 7-4. *Cluster Centers*

	Cluster 1	Cluster 2	Cluster 3
0	-1.600930	9.258221	-4.160474
1	-1.429737	0.508989	2.217940
2	-0.587821	0.273357	0.862871

Listing 7-12 plots Figure 7-4, which shows the estimated labels.

Listing 7-12. Africa's GNI Per Capita Scatter

```
fig, ax = plt.subplots()
plt.scatter(pca_df[:,0],pca_df[:,1],c=kmeans_output.
labels_,cmap="viridis",s=200)
plt.scatter(centers[:,0], centers[:,1], color="red")
plt.title("Africa GNI per capita, Atlas method (current US$) clusters")
plt.xlabel("y")
plt.show()
```

Africa GNI per capita, Atlas method (current US$) clusters

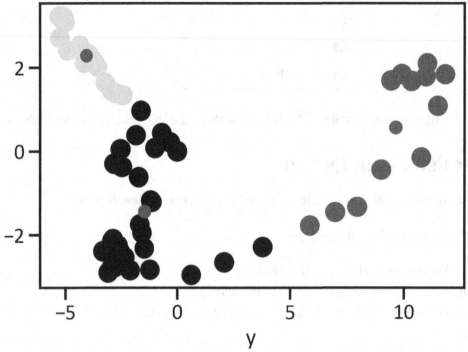

Figure 7-4. *Africa GNI per capita k-means model*

Figure 7-4 shows the data points (highlighted in yellow, purple, and green) in their respective clusters (red dots).

Cluster Results Analysis

To better comprehend the predicted labels from Listing 7-13, see Table 7-5.

Listing 7-13. Africa GNI Per Capita Cluster Table

```
stocks = pd.DataFrame(df.columns)
cluster_labels = pd.DataFrame(kmeans.labels_)
stockClusters = pd.concat([stocks, cluster_labels],axis = 1)
stockClusters.columns = ["Country","Clusters"]
stockClusters = stockClusters.replace({0: "Low GNI per capita",1: "Moderate
GNI per capita",2: "High GNI per capita"})
stockClusters.head(20)
```

Table 7-5. *Africa GNI Per Capita Cluster Table*

	Country	Clusters
0	Algeria	Low GNI per capita
1	Angola	Low GNI per capita
2	Benin	High GNI per capita
3	Botswana	High GNI per capita
4	Burundi	High GNI per capita
5	Central African Republic	High GNI per capita
6	Chad	High GNI per capita
7	Congo, Dem. Rep.	High GNI per capita
8	Cote d'Ivoire	High GNI per capita
9	Egypt, Arab Rep.	High GNI per capita
10	Equatorial Guinea	High GNI per capita
11	Eswatini	High GNI per capita
12	Ethiopia	High GNI per capita
13	Gabon	High GNI per capita

(continued)

Table 7-5. (*continued*)

	Country	Clusters
14	Gambia, The	High GNI per capita
15	Ghana	High GNI per capita
16	Guinea	High GNI per capita
17	Kenya	High GNI per capita
18	Lesotho	High GNI per capita
19	Liberia	Low GNI per capita

Listing 7-14 simplifies the data in Table 7-5 (see Figure 7-5).

Listing 7-14. Africa's GNI Per Capita Cluster Count

```
class_series = stockClusters.groupby("Clusters").size()
class_series.name = "Count"
class_series.plot.pie(autopct="%2.f",cmap="Blues_r")
plt.title("Clusters pie chart")
plt.show()
```

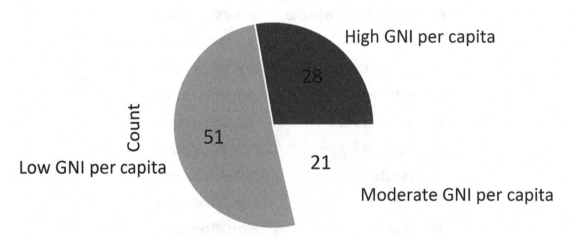

Figure 7-5. *Africa's GNI per capita cluster count*

Figure 7-5 shows that 28% of the countries have a high GNI per capita, 51% have a low GNI per capita, and 21% have a moderate GNI per capita.

K-Means Model Evaluation

The k-means model does not make extreme assumptions about the data. You can apply it with no predetermined hypothesis—it establishes truth about the data rather than testing claims. It also does not have robust model evaluation matrices.

The Silhouette Methods

You can depend on the Silhouette method to determine the extent to which the model makes intelligent guesstimates (see Listing 7-15). It estimates the difference between the mean nearest-cluster distance and the intra-cluster distance and the maximum mean nearest-cluster distance and the intra-cluster distance. It returns values that range from -1 to 1, where -1 indicates a poor model, 0 indicates a mediocre model, and 1 indicates a model that's exemplary in making guesstimates.

Listing 7-15. Find Silhouette Score

```
from sklearn import metrics
metrics.silhouette_score(df, y_predkmeans)
0.43409785393807543
```

The silhouette score is 0.43401. This means that the classifier is mediocre in estimating labels of GNI per capita for the African countries.

Conclusion

This chapter introduced an unsupervised machine learning model for clustering, called the *k-means model*. It developed a model with default hyperparameters. It estimated the distance between the values of GNI per capita in Africa's countries. Afterward, it grouped the countries into clusters by GNI per capita. This chapter concludes with the discussion of unsupervised machine learning.

CHAPTER 8

Solving Economic Problems Applying Artificial Neural Networks

Deep learning extends machine learning by using artificial neural networks to operate a set of predictor variables and predict future response variables. Artificial neural networks are a group of nodes that receive input values in the input layer and transform them in the subsequent hidden layer (a layer in between the input and output layer). This hidden layer transforms the nodes and allots varying *weights* (vector parameters that determine the extent of influence input values have on output values) and *bias* (a balance value which invariably is 1). Following that, they generate a set of output values in the output layer by applying an activation function. Artificial neural networks are part of deep learning, which advances machine learning through structuring models and by replicating human neural activity as the base. *Propagation* is the process of training networks (often through backward propagation, which involves updating weights in reserves, from the output layer).

Traditional machine learning models are most suitable to small data. Increasing the volume of data presents fresh problems, as conventional models are computationally expensive. Most ensemble models (those that solve regression and classification problems) are excessively optimistic in the beginning phases of training and get most of the predictions correct. For instance, a gradient seems small at first and then expands as you increase data—this phenomenon is called the *vanishing gradient problem*. The gradient descent algorithm contends with this problem, which forms the basis for structuring the network.

T. C. Nokeri, *Econometrics and Data Science*, https://doi.org/10.1007/978-1-4842-7434-7_8

Before you proceed, be sure that you have the `tensorflow` library installed in your environment. To install the `tensorflow` library in a Python environment, use `pip install tensorflow`. Equally, to install the library in a Conda environment, use `conda install -c conda-forge tensorflow`. Install the Keras library as well. On a Python environment, use `pip install keras` and on the Conda environment, use `conda install -c conda-forge keras`.

Context of This Chapter

This chapter predicts changes in life expectancy at birth in South Africa by applying a set of economic indicators—urban population, GNI per capita, Atlas method (in current U.S. dollars), and GDP growth (as an annual percentage). Table 8-1 outlines South Africa's macroeconomic indicators that this chapter investigates.

Table 8-1. *South Africa's Macroeconomic Indicators for This Chapter*

ID	Indicators
SP.URB.TOTL	South Africa's urban population
NY.GNP.PCAP.CD	South Africa's GNI per capita, Atlas method (in current U.S. dollars)
NY.GDP.MKTP.KD.ZG	South Africa's GDP growth (as an annual percentage)
SP.DYN.LE00.IN	South Africa's life expectancy at birth (in years)

Theoretical Framework

Figure 8-1 illustrates the hypothetical framework for this chapter.

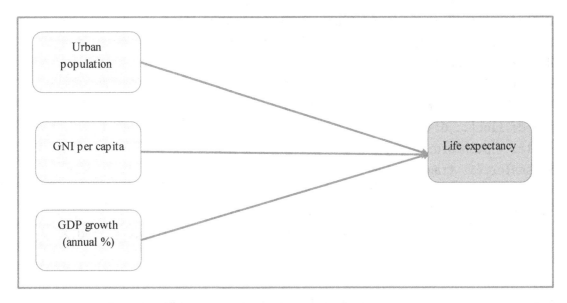

Figure 8-1. *Theoretical framework*

The code in Listing 8-1 performs the data manipulation tasks.

Listing 8-1. Data Preprocessing

```
country = ["ZAF"]
indicator = {"SP.URB.TOTL":"urban_pop",
             "NY.GNP.PCAP.CD":"gni_per_capita",
             "NY.GDP.MKTP.KD.ZG":"gdp_growth",
             "SP.DYN.LE00.IN":"life_exp"}
df = wbdata.get_dataframe(indicator, country=country, convert_date=True)
df["urban_pop"] = df["urban_pop"].fillna(df["urban_pop"].mean())
df["gni_per_capita"] = df["gni_per_capita"].fillna(df["gni_per_capita"].mean())
df["gdp_growth"] = df["gdp_growth"].fillna(df["gdp_growth"].mean())
df["life_exp"] = df["life_exp"].fillna(df["life_exp"].mean())
df['gni_per_capita'] = np.where((df["gni_per_capita"] > 5000),
df["gni_per_capita"].mean(),df["gni_per_capita"])
df['pct_change'] = df["life_exp"].pct_change()
df = df.dropna()
df['direction'] = np.sign(df['pct_change']).astype(int)
df["direction"] = pd.get_dummies(df["direction"])
df['pct_change'] = df["life_exp"].pct_change()
df = df.dropna()
```

```
df['direction'] = np.sign(df['pct_change']).astype(int)
df["direction"] = pd.get_dummies(df["direction"])
del df["pct_change"]
del df["life_exp"]
from sklearn.preprocessing import StandardScaler
x = df.iloc[::,0:3]
scaler = StandardScaler()
x= scaler.fit_transform(x)
y = df.iloc[::,-1]
from sklearn.model_selection import train_test_split
x_train, x_test, y_train, y_test = train_test_split(x,y,test_size=0.2,
random_state=0)
```

Restricted Boltzmann Machine Classifier

The restricted Boltzmann machine classifier is a simple artificial neural network that consists of one hidden layer and one visible layer. Figure 8-2 shows the structure of a restricted Boltzmann machine classifier.

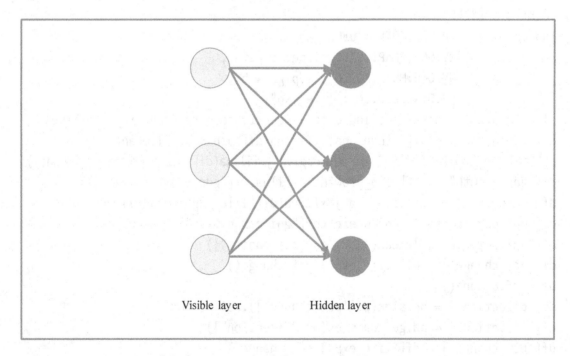

Figure 8-2. *Restricted Boltzmann machine structure*

Figure 8-2 shows that each node in the hidden layer automatically retrieves input values, transforms the retrieved input data by applying an activation function, and transmits the data to the visible layer. The restricted Boltzmann machine classifier estimates the gradient from the left to the right—this learning process is called the *forward propagation* process.

Restricted Boltzmann Machine Classifier Development

Listing 8-2 is the Restricted Boltzmann Machine Development classifier, applying default hyperparameters. Notice that it applies the `Pipeline()` method to set parameters of the final classifier, including those of the restricted Boltzmann machine and logistic regression classifier.

Listing 8-2. Restricted Boltzmann Machine Classifier Development

```
from sklearn.neural_network import BernoulliRBM
from sklearn.linear_model import LogisticRegression
rbm = BernoulliRBM()
logreg = LogisticRegression()
from sklearn.pipeline import Pipeline
rbmclassifier = Pipeline(steps=[("rbm",rbm),("logreg",logreg)])
rbmclassifier.fit(x_train,y_train)
```

Restricted Boltzmann Machine Confusion Matrix

Listing 8-3 investigates the restricted Boltzmann machine classifier's performance (see Table 8-2).

Listing 8-3. Restricted Boltzmann Machine Confusion Matrix

```
from sklearn import metrics
y_predrbmclassifier = rbmclassifier.predict(x_test)
cmatrbmclassifier = pd.DataFrame(metrics.confusion_matrix(y_test,
y_predrbmclassifier),
                           index=["Actual: Decreasing life expectancy",
```

```
                              "Actual: Increasing life expectancy"],
                 columns = ("Predicted: Decreasing life expectancy",
                            Predicted: Increasing life expectancy"))
cmatrbmclassifier
```

Table 8-2. *Restricted Boltzmann Machine Classifier Confusion Matrix*

	Predicted: Decreasing Life Expectancy	Predicted: Increasing Life Expectancy
Actual: Decreasing Life Expectancy	0	2
Actual: Increasing Life Expectancy	0	10

Table 8-3 interprets the restricted Boltzmann machine classifier's confusion matrix (see Table 8-2).

Table 8-3. *Restricted Boltzmann Machine Classifier Confusion Matrix Interpretation*

Metric	Interpretation
True Positive	The restricted Boltzmann machine classifier did not predict decreasing life expectancy when there was an actual decreasing life expectancy in South Africa.
True Negative	The restricted Boltzmann machine classifier predicted increasing life expectancy when there was increasing life expectancy in South Africa (two times).
True Positive	The restricted Boltzmann machine classifier did not predict decreasing life expectancy when there was increasing life expectancy in South Africa.
False Negative	The restricted Boltzmann machine classifier predicted increasing life when there actually was decreasing life expectancy in South Africa (ten times).

Restricted Boltzmann Machine Classification Report

The code in Listing 8-4 retrieves a detailed report regarding the performance of the restricted Boltzmann machine classifier (see Table 8-4).

Listing 8-4. Restricted Boltzmann Machine Classification Report

```
creportrbmclassifier = pd.DataFrame(metrics.classification_report(y_test,
y_predrbmclassifier,output_dict=True))
creportrbmclassifier
```

Table 8-4 shows that the RBM classifier is accurate 83% of the time, and it is 83% precise when predicting increasing and decreasing life expectancy in South Africa.

Table 8-4. *Restricted Boltzmann Machine Classification Report*

	0	1	Accuracy	Macro Avg	Weighted Avg
Precision	0.0	0.833333	0.833333	0.416667	0.694444
Recall	0.0	1.000000	0.833333	0.500000	0.833333
F1-score	0.0	0.909091	0.833333	0.454545	0.757576
Support	2.0	10.000000	0.833333	12.000000	12.000000

Restricted Boltzmann Machine Classifier ROC Curve

Listing 8-5 estimates the trade-offs between sensitivity (FPR) and specificity (TPR) as the restricted Boltzmann machine classifier distinguishes between increasing and decreasing life expectancy in South Africa (see Figure 8-3).

Listing 8-5. Restricted Boltzmann Machine Classifier ROC Curve

```
y_predrbmclassifier_proba = rbmclassifier.predict_proba(x_test)[::,1]
fprrbmclassifier, tprrbmclassifier, _ = metrics.roc_curve(y_test,
y_predrbmclassifier_proba)
aucrbmclassifier = metrics.roc_auc_score(y_test,y_predrbmclassifier_proba)
fig, ax = plt.subplots()
plt.plot(fprrbmclassifier, tprrbmclassifier,label="auc: "+
str(aucrbmclassifier),color="navy",lw=4)
plt.plot([0,1],[0,1],color="red",lw=4)
plt.xlim([0.00,1.01])
plt.ylim([0.00,1.01])
plt.title("RBM classifier ROC curve")
plt.xlabel("Sensitivity")
```

```
plt.ylabel("Specificity")
plt.legend(loc=4)
plt.show()
```

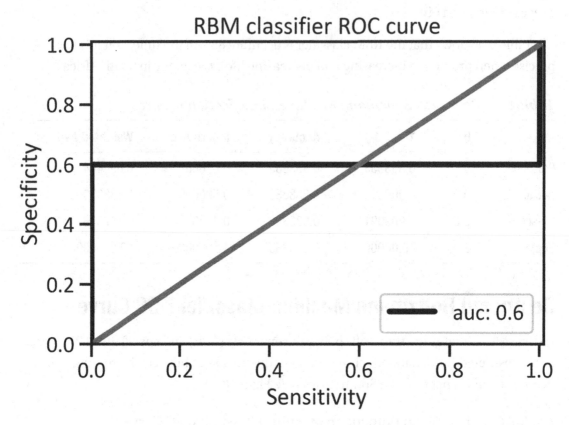

Figure 8-3. *Restricted Boltzmann machine classifier ROC curve*

Figure 8-3 shows that, on average, the restricted Boltzmann machine classifier is 60% accurate when distinguishing between increasing and decreasing life expectancy at birth in South Africa.

Restricted Boltzmann Machine Classifier Precision-Recall Curve

Listing 8-6 estimates the trade-off between precision and recall across certain thresholds as the restricted Boltzmann machine classifier distinguishes between increasing and decreasing life expectancy in South Africa (see Figure 8-4).

Listing 8-6. Restricted Boltzmann Machine Classifier Precision-Recall Curve

```
precisionrbmclassifier, recallrbmclassifier, thresholdrbmclassifier =
metrics.precision_recall_curve(y_test,y_predrbmclassifier)
apsrbmclassifier = metrics.average_precision_score(y_test,
y_predrbmclassifier)
fig, ax = plt.subplots()
plt.plot(precisionrbmclassifier, recallrbmclassifier,label="aps: " +
str(apsrbmclassifier),color="navy",lw=4)
plt.axhline(y=0.5, color="red",lw=4)
plt.title("RBM classifier precision-recall curve")
plt.ylabel("Precision")
plt.xlabel("Recall")
plt.legend(loc=4)
plt.show()
```

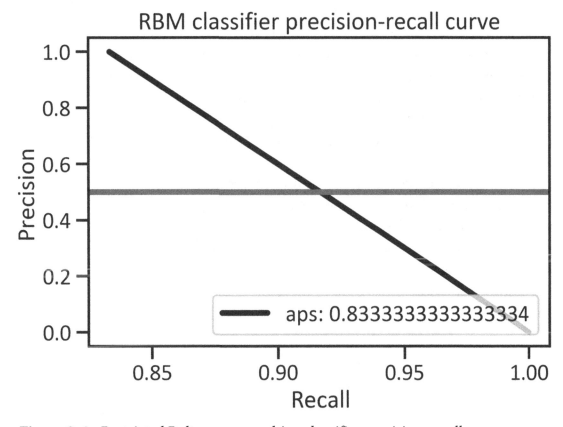

Figure 8-4. *Restricted Boltzmann machine classifier precision-recall curve*

Figure 8-4 shows that, on average, the restricted Boltzmann machine classifier is 83% precise when distinguishing between increasing and decreasing life expectancy in South Africa.

Restricted Boltzmann Machine Classifier Learning Curve

Listing 8-7 demonstrates how the restricted Boltzmann machine classifier learns to correctly distinguish between increasing and decreasing life expectancy at birth in South Africa (see Figure 8-5).

Listing 8-7. Restricted Boltzmann Machine Classifier Learning Curve

```
from sklearn.model_selection import learning_curve
from sklearn.model_selection import learning_curve
trainsizerbmclassifier, trainscorerbmclassifier, testscorerbmclassifier =
learning_curve(rbmclassifier, x, y, cv=5, n_jobs=-1, scoring="accuracy",
train_sizes=np.linspace(0.1,1.0,50))
trainscorerbmclassifier_mean = np.mean(trainscorerbmclassifier,axis=1)
trainscorerbmclassifier_std = np.std(trainscorerbmclassifier,axis=1)
testscorerbmclassifier_mean = np.mean(testscorerbmclassifier,axis=1)
testscorerbmclassifier_std = np.std(testscorerbmclassifier,axis=1)
fig, ax = plt.subplots()
plt.plot(trainsizerbmclassifier,trainscorerbmclassifier_mean,color="red",label="Training
accuracy score",lw=4)
plt.plot(trainsizerbmclassifier, testscorerbmclassifier_
mean,color="navy",label="Cross validation accuracy score",lw=4)
plt.title("RBM classifier learning curve")
plt.xlabel("Training set size")
plt.ylabel("Accuracy")
plt.legend(loc="best")
plt.show()
```

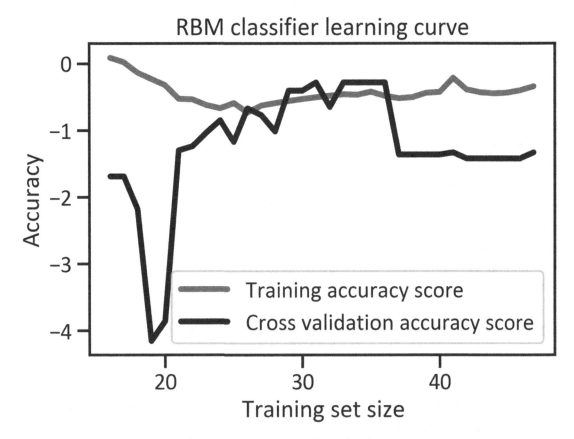

Figure 8-5. *Restricted Boltzmann machine classifier learning curve*

Figure 8-5 shows that, for the first ten data points, the cross-validation accuracy score is higher than the training score. Thereafter, the training score surpasses the cross-validation accuracy score up to the 30th data point. At the 36th data point, the cross-validation score sharply declines.

Multilayer Perceptron (MLP) Classifier

This section covers the Multilayer Perceptron classifier—an extension of the restricted Boltzmann machine classifier, also called the *vanilla model*. It is a restricted Boltzmann machine classifier with more than one hidden layer. It comprises three layers (see Figure 8-6).

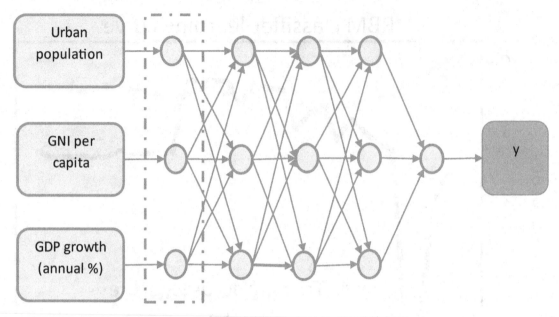

Figure 8-6. *Example of a Multilayer Perceptron classifier*

Figure 8-6 shows that the model includes the input layer, which retrieves input values and transmits them to the first hidden layer. That hidden layer receives the values and transforms them by applying a function (in this case, the Sigmoid function). The output value is transmitted to the second hidden layer, which receives the transmitted values and processes them (it transforms the values and transmits them to the output layer.).

Multilayer Perceptron (MLP) Classifier Model Development

Listing 8-8 develops the Multilayer Perceptron classifier by applying default hyper-parameters.

Listing 8-8. Multilayer Perceptron Classifier Model Development

```
from sklearn.neural_network import MLPClassifier
mlp = MLPClassifier()
mlp.fit(x_train,y_train)
```

Listing 8-9 retrieves the Multilayer Perceptron classifier's confusion matrix (see Table 8-5).

Listing 8-9. Construct a Multilayer Perceptron Confusion Matrix

```
y_predmlp = mlp.predict(x_test)
cmatmlp = pd.DataFrame(metrics.confusion_matrix(y_test,y_predmlp),
index=["Actual: Decreasing life expectancy", "Actual: Increasing life
expectancy"],
                          columns = ("Predicted: Decreasing life expectancy",
                                "Predicted: Increasing life expectancy"))
cmatmlp
```

Table 8-5. *Multilayer Perceptron Confusion Matrix*

	Predicted: Decreasing Life Expectancy	Predicted: Increasing Life Expectancy
Actual: Decreasing Life Expectancy	0	2
Actual: Increasing Life Expectancy	0	10

Table 8-6 interprets the Multilayer Perceptron classifier's confusion matrix (see Table 8-5).

Table 8-6. *Multilayer Perceptron Confusion Matrix Interpretation*

Metric	Interpretation
True Positive	The Multilayer Perceptron classifier did not predict decreasing life expectancy, when there actually was decreasing life expectancy in South Africa.
True Negative	The Multilayer Perceptron classifier predicted increasing life expectancy at birth, when there was actually increasing life expectancy in South Africa (two times).
True Positive	The Multilayer Perceptron classifier did not predict decreasing life expectancy, when there was actually increasing life expectancy at birth in South Africa.
False Negative	The Multilayer Perceptron classifier predicted increasing life expectancy, when there was actually decreasing life expectancy in South Africa (ten times).

Multilayer Perceptron Classification Report

Table 8-7 outlines the Multilayer Perceptron classification report created by the code in Listing 8-10.

Listing 8-10. Multilayer Perceptron Classification Report

```
creportmlp = pd.DataFrame(metrics.classification_report(y_test,
y_predmlp,output_dict=True))
creportmlp
```

Table 8-7. *Multilayer Perceptron Classification Report*

	0	1	Accuracy	Macro Avg	Weighted Avg
Precision	0.0	0.833333	0.833333	0.416667	0.694444
Recall	0.0	1.000000	0.833333	0.500000	0.833333
F1-score	0.0	0.909091	0.833333	0.454545	0.757576
Support	2.0	10.000000	0.833333	12.000000	12.000000

Table 8-7 shows that the Multilayer Perceptron classifier is accurate 83% of the time when it predicts decreasing life expectancy at birth in South Africa, and it is precise 83% of the time when it predicts decreasing life expectancy. In addition, it is accurate 92% of the time when it distinguishes between increasing and decreasing life expectancy in South Africa.

Multilayer Perceptron ROC Curve

Listing 8-11 estimates the trade-offs between sensitivity (FPR) and specificity (TPR) as the Multilayer Perceptron classifier distinguishes between increasing and decreasing life expectancy in South Africa (see Figure 8-7).

Listing 8-11. Multilayer Perceptron ROC Curve

```
y_predmlp_proba = mlp.predict_proba(x_test)[::,1]
fprmlp, tprmlp, _ = metrics.roc_curve(y_test,y_predmlp_proba)
aucmlp = metrics.roc_auc_score(y_test,y_predmlp_proba)
fig, ax = plt.subplots()
```

```
plt.plot(fprmlp, tprmlp,label="auc: "+str(aucmlp),color="navy",lw=4)
plt.plot([0,1],[0,1],color="red",lw=4)
plt.xlim([0.00,1.01])
plt.ylim([0.00,1.01])
plt.title("MLP classifier ROC curve")
plt.xlabel("Sensitivity")
plt.ylabel("Specificity")
plt.legend(loc=4)
plt.show()
```

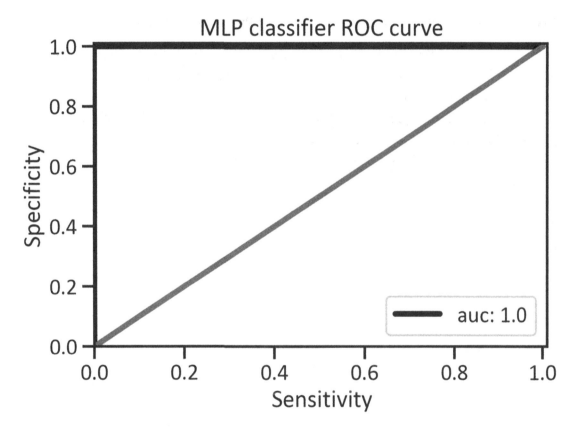

Figure 8-7. *Multilayer Perceptron classifier ROC curve*

Figure 8-7 shows that on average, the Multilayer Perceptron classifier is 100% accurate when distinguishing between increasing and decreasing life expectancy in South Africa.

Multilayer Perceptron Classifier Precision-Recall Curve

Listing 8-12 estimates the trade-off between precision and recall across various thresholds as the multilayer perceptron classifier distinguishes between increasing and decreasing life expectancy in South Africa (see Figure 8-8).

Listing 8-12. Multilayer Perceptron Classifier Precision Recall Curve

```
precisionmlp,recallmlp,thresholdmlp=metrics.precision_recall_curve(y_test,y_predmlp)
apsmlp = metrics.average_precision_score(y_test,y_predmlp)
fig, ax = plt.subplots()
plt.plot(precisionmlp,recallmlp,label="aps: "+str(apsmlp),color="navy",lw=4)
plt.axhline(y=0.5, color="red",lw=4)
plt.title("MLP classifier precision-recall curve")
plt.ylabel("Precision")
plt.xlabel("Recall")
plt.legend(loc=4)
plt.show()
```

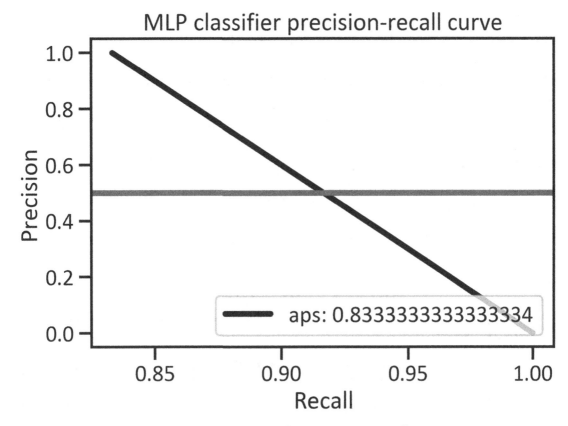

Figure 8-8. *Multilayer Perceptron classifier precision-recall curve*

Figure 8-8 demonstrates that on average, the Multilayer Perceptron classifier is 83% precise when distinguishing between increasing and decreasing life expectancy in South Africa.

Multilayer Perceptron Classifier Learning Curve

Listing 8-13 demonstrates how the Multilayer Perceptron classifier learns to correctly distinguish between increasing and decreasing life expectancy at birth in South Africa as the training test size increases (see Figure 8-9).

Listing 8-13. MLP Classifier Learning Curve

```
trainsizemlp, trainscoremlp, testscoremlp = learning_curve(mlp, x, y, cv=5,
n_jobs=-1, scoring="accuracy", train_sizes=np.linspace(0.1,1.0,50))
trainscoremlp_mean = np.mean(trainscoremlp,axis=1)
trainscoremlp_std = np.std(trainscoremlp,axis=1)
```

```
testscoremlp_mean = np.mean(testscoremlp,axis=1)
testscoremlp_std = np.std(testscoremlp,axis=1)
fig, ax = plt.subplots()
plt.plot(trainsizemlp, trainscoremlp_mean, color="red",lw=4,label="Training
accuracy score")
plt.plot(trainsizemlp, testscoremlp_mean,color="navy",lw=4,label="Cross
validation accuracy score")
plt.title("MLP classifier learning curve")
plt.xlabel("Training set size")
plt.ylabel("Accuracy")
plt.legend(loc="best")
plt.show()
```

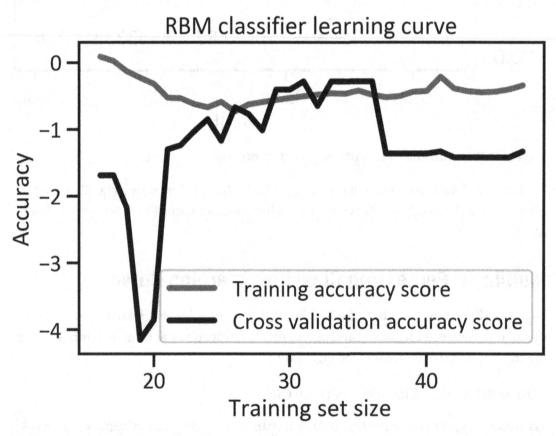

Figure 8-9. *Multilayer Perceptron classifier learning curve*

Figure 8-9 shows that the training accuracy score is higher than the cross-validation accuracy score throughout the training process. The cross-validation score sharply declines after the 12th data point and increases at the 20th data point.

Artificial Neural Network Prototyping Using Keras

Listing 8-14 reprocesses the data. This procedure is like the one applied in the preceding section. The difference is that this process further breaks down the data into validation data.

Listing 8-14. Data Reprocessing

```
from sklearn.preprocessing import StandardScaler
x = df.iloc[::,0:3]
scaler = StandardScaler()
x= scaler.fit_transform(x)
y = df.iloc[::,-1]
from sklearn.model_selection import train_test_split
x_train, x_test, y_train, y_test = train_test_split(x,y,test_size=0.2,
random_state=0)
x_train, x_val, y_train, y_val = train_test_split(x_train,y_train,
test_size=0.1,random_state=0)
```

Listing 8-15 trains the neural network. It begins by importing Sequential, Dense and KerasClassifier from the keras library (see Listing 8-15).

Listing 8-15. Import Dependencies

```
from keras import Sequential
from keras.layers import Dense
from keras.wrappers.scikit_learn import KerasClassifier
```

Artificial Neural Network Structuring

Listing 8-16 builds a function to create the architecture of the neural network. The activation function in the input layer is ReLu. This means that the model will operate on the variables—urban population, GNI per capita, Atlas method (in current U.S. dollars), and GDP growth (as an annual percentage)—and generate output. Listing 8-16 includes an extra node in the input layer. In the output layer, the network applies the sigmoid (it operates on these predictor variables and generates decreasing or increasing life expectancy) in South Africa. The loss function applied to the training is binary_crossentropy. Specifying this loss function informs the neural network that it will deal with a categorical variable. The metric applied to evaluate the neural network is accuracy and we apply the adam optimizer to select variables at random. We also approach the vanishing gradient problem and quicken the training process.

Listing 8-16. Build Network Architecture

```
def create_dnn_model():
    model = Sequential()
    model.add(Dense(4, input_dim=3, activation="relu"))
    model.add(Dense(1, activation="sigmoid"))
    model.compile(loss="binary_crossentropy", optimizer="adam",
    metrics=["accuracy"])
    return model
```

Network Wrapping

Listing 8-17 wraps the classifier, so we apply the scikit-learn library functionalities.

Listing 8-17. Wrap Keras Classifier

```
model = KerasClassifier(build_fn=create_dnn_model)
```

Listing 8-18 trains the neural network on 64 epochs, applying a batch size of 15. Note that you can fiddle with the settings by trying other batch sizes and epochs (they frequently give different model performance results).

Listing 8-18. Train Neural Network

```
history = model.fit(x_train, y_train, validation_data=(x_val, y_val),
epochs=64, batch_size=15)
history
```

Keras Classifier Confusion Matrix

Listing 8-19 creates a confusion matrix (see Table 8-8).

Listing 8-19. Keras Classifier Confusion Matrix

```
y_predkeras = model.predict(x_test)
cmatkeras = pd.DataFrame(metrics.confusion_matrix(y_test,y_predkeras),
                    index=["Actual: Decreasing life expectancy",
                          "Actual: Increasing life expectancy"],
                    columns = ("Predicted: Decreasing life expectancy",
                          "Predicted: Increasing life expectancy"))
cmatkeras
```

Table 8-8. *Keras Classifier Confusion Matrix*

	Predicted: Decreasing Life Expectancy	Predicted: Increasing Life Expectancy
Actual: Decreasing Life Expectancy	0	2
Actual: Increasing Life Expectancy	1	9

Table 8-9 interprets the Keras classifier's confusion matrix (see Table 8-8).

Table 8-9. *Keras Classifier Confusion Matrix Interpretation*

Metric	Interpretation
True Positive	The Keras classifier did not predict decreasing life expectancy when there was actually decreasing life expectancy in South Africa times.
True Negative	The Keras classifier predicted increasing life expectancy when there was actually increasing life expectancy in South Africa (two times).
True Positive	The Keras classifier predicted decreasing life expectancy when there was actually increasing life expectancy in South Africa (one time).
False Negative	The Keras classifier predicted increasing life expectancy when there was actually decreasing life expectancy in South Africa (nine times).

Keras Classification Report

Listing 8-20 generates a classification report (see Table 8-10).

Listing 8-20. Keras Classification Report

```
creportkeras = pd.DataFrame(metrics.classification_report(y_test,
y_predkeras,output_dict=True))
creportkeras
```

Table 8-10. *Keras Classification Report*

	0	1	Accuracy	Macro Avg	Weighted Avg
Precision	0.0	0.818182	0.75	0.409091	0.681818
Recall	0.0	0.900000	0.75	0.450000	0.750000
F1-score	0.0	0.857143	0.75	0.428571	0.714286
Support	2.0	10.000000	0.75	12.000000	12.000000

Table 8-10 shows that the Keras classifier is accurate 41.33% of the time when it predicts decreasing life expectancy in South Africa, and it is precise 81.0% of the time when it predicts decreasing life expectancy. In addition, it is accurate 0% of the time when it distinguishes between increasing and decreasing life expectancy.

Keras Classifier ROC Curve

Listing 8-21 determines the trade-off between precision and recall across various thresholds as the Keras classifier distinguishes between increasing and decreasing life expectancy in South Africa (see Figure 8-10).

Listing 8-21. Keras Classifier ROC Curve

```
y_predkeras_proba = model.predict_proba(x_test)[::,1]
fprkeras, tprkeras, _ = metrics.roc_curve(y_test,y_predkeras_proba)
auckeras = metrics.roc_auc_score(y_test,y_predkeras_proba)
fig, ax = plt.subplots()
plt.plot(fprkeras, tprkeras,label="auc: "+str(auckeras),color="navy",lw=4)
```

```
plt.plot([0,1],[0,1],color="red",lw=4)
plt.xlim([0.00,1.01])
plt.ylim([0.00,1.01])
plt.title("Keras classifier ROC curve")
plt.xlabel("Sensitivity")
plt.ylabel("Specificity")
plt.legend(loc=4)
plt.show()
```

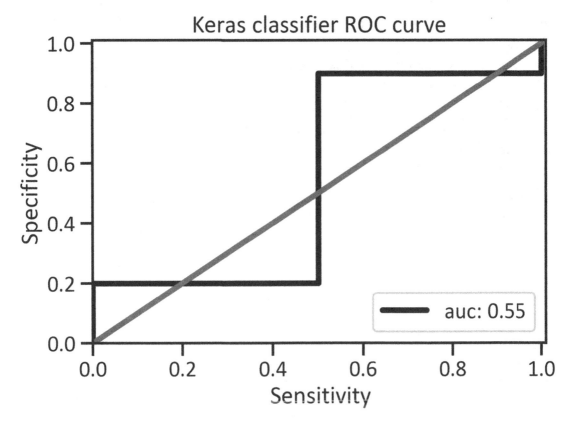

Figure 8-10. *Keras classifier ROC curve*

Figure 8-10 shows that on average, the Keras classifier is 55% accurate when distinguishing between increasing and decreasing life expectancy in South Africa.

Keras Classifier Precision-Recall Curve

Listing 8-22 estimates the trade-off between precision and recall across various thresholds as the Keras classifier distinguishes between increasing and decreasing life expectancy in South Africa (see Figure 8-11).

Listing 8-22. Keras Classifier Precision-Recall Curve

```
precisionkeras, recallkeras, thresholdkeras = metrics.precision_recall_
curve(y_test,y_predkeras)
apskeras = metrics.average_precision_score(y_test,y_predkeras)
fig, ax = plt.subplots()
plt.plot(precisionkeras, recallkeras,label="aps: " +str(apskeras),
color="navy",lw=4)
plt.axhline(y=0.5, color="red",lw=4)
plt.title("Keras classifier precision-recall curve")
plt.ylabel("Precision")
plt.xlabel("Recall")
plt.legend(loc=4)
plt.show()
```

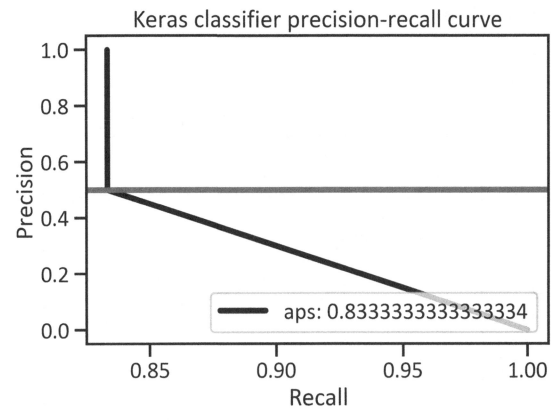

Figure 8-11. *Keras classifier precision-recall curve*

Figure 8-11 shows that, on average, the Keras classifier is 83% precise when distinguishing between increasing and decreasing life expectancy in South Africa.

Training Loss and Cross-Validation Loss Across Epochs

Listing 8-23 determines the loss across epochs, whereby epochs is in the x-axis and loss is in the y-axis (see Figure 8-12).

Listing 8-23. Training Loss and Cross-Validation Loss Across Epochs

```
plt.plot(history.history["loss"], color="red",lw=4, label="Training loss")
plt.plot(history.history["val_loss"], color="navy",lw=4, label="Cross
validation loss")
plt.title("Training loss and cross validation loss across epochs")
```

```
plt.xlabel("Loss")
plt.ylabel("Epochs")
plt.legend(loc="best")
plt.show()
```

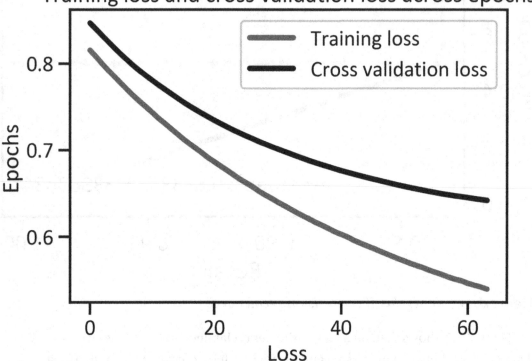

Figure 8-12. *Training loss and cross-validation loss across epochs*

Figure 8-12 demonstrates that, starting with the first epoch, both the training loss and cross-validation loss gradually decline until they reach the lowest point at the 65th epoch.

Training Loss and Cross-Validation Loss Accuracy Across Epochs

Listing 8-24 estimates the accuracy across epochs (see Figure 8-13).

Listing 8-24. Training Loss and Cross-Validation Loss Accuracy Epochs

```
plt.plot(history.history["accuracy"], color="red",lw=4, label="Training
accuracy")
plt.plot(history.history["val_accuracy"], color="navy",lw=4,label="Cross
validation accuracy")
plt.title("Training loss and cross validation loss accuracy epochs")
plt.xlabel("Loss")
plt.ylabel("Epochs")
plt.legend(loc="best")
plt.show()
```

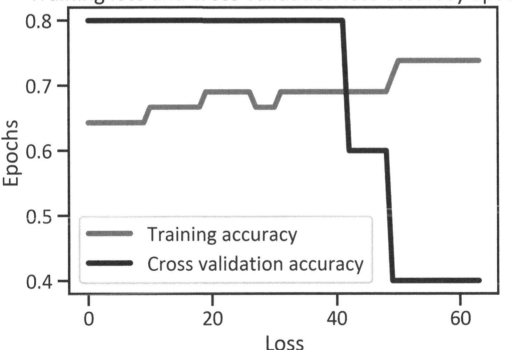

Figure 8-13. *Training loss and cross-validation loss accuracy epochs*

Figure 8-13 demonstrates that from the first epoch, the cross-validation accuracy is higher than the training accuracy. Moreover, before the 45th epoch, the cross-validation accuracy sharply declines, then becomes lower than the training accuracy

Conclusion

This chapter covered deep learning. It unpacked fundamental deep learning concepts. It introduced techniques for developing artificial neural networks using the `scikit-learn` library. You learned how to develop and evaluate a shallow artificial neural network called the restricted Boltzmann machine, including a Multilayer Perceptron. Based on the findings, the Multilayer Perceptron classifier that was developed by applying the `scikit-learn` library outperformed all the networks. In addition, it showed ways of developing a fundamental network architecture for a sequential model by applying the `keras` library.

CHAPTER 9

Inflation Simulation

This chapter investigates the impact of different scenarios on Great Britain's inflation and consumer prices (as an annual percentage) using the Monte Carlo simulation model. In particular, it employs this model to determine the probability of a change in the country's central government debt across multiple trials. This method is useful when handling sequential data.

Understanding Simulation

This chapter considers the Monte Carlo simulation model. It's an unsupervised model, given that it does not split the data—the model explores the entire data set. It learns the entire data across multiple trials to account for the possibility of an event occurring. In addition, it investigates the impact of different scenarios of Great Britain's central government debt by applying the Monte Carlo model—which determines the probability of a change in central government debt across ten scenarios (the number of simulations). This model is used in finance scenarios to determine risks and uncertainty relating to investing in a certain asset class.

This method reproduces values in variables across varying iterations. By doing so, it enables you to examine changes in the data numerous times. There are many simulation methods (i.e., agent-based simulation, discrete event simulation, system dynamics simulation, and Monte Carlo simulation). This chapter only acquaints you with the most common simulation method—the Monte Carlo simulation method. It simulates changes and recognizes patterns in past occurrences and forecast values of a sequence, thus enabling developers to replicate events in the real world.

Unlike models covered in the preceding chapters (besides k-means), this model does not require you to split the data. In addition to that, it enables repetitive examination of data, thus providing a clear picture of changes. Moreover, it does not contain sophisticated metrics for evaluation; the best you can do is the max drawdown

© Tshepo Chris Nokeri 2022
T. C. Nokeri, *Econometrics and Data Science*, https://doi.org/10.1007/978-1-4842-7434-7_9

(extreme negative peaks), including the mean and standard deviation of the simulation results. Using simulation methods is widespread in finance for understanding risk without exposing oneself (learn more here[1]).

This model is suitable for policymakers, as it can help them have clarity when considering uncertainty in economic events. It can help them decide when to devise and revise policies that impact borrowing activities. It is very intuitive, and you apply it to a variable across multiple trials to generate output (a simulation).

Before you proceed, ensure you have the `pandas_montecarlo` library installed in your environment. To install the `pandas_montecarlo` library in a Python environment, use `pip install pandas-montecarlo`. Equally, to install the library in a Conda environment, use `conda install pandas_montecarlo`.

Context of This Chapter

This chapter shows how to predict Great Britain's inflation and consumer prices (as an annual percentage) across multiple trials by applying the Monte Carlo model. Table 9-1 outlines the macroeconomic indicator that this chapter investigates.

Table 9-1. *A Great Britain's Indicator for This Chapter*

ID1	Indicator
FP.CPI.TOTL.ZG	Inflation and consumer prices (as an annual percentage)

Listing 9-1 extracts Great Britain's inflation and consumer prices data (see Table 9-2).

Listing 9-1. Great Britain's Inflation and Consumer Prices

```
import wbdata
country  = ["GBR"]
indicator = {"FP.CPI.TOTL.ZG":"inflation_cpi"}
df = wbdata.get_dataframe(indicator, country=country,convert_date=True)
df.head()
```

---

[1] https://link.springer.com/chapter/10.1007/978-1-4842-7110-0_4

Table 9-2. *Great Britain's Inflation and Consumer Prices*

Date	inflation_cpi
2020-01-01	0.989487
2019-01-01	1.738105
2018-01-01	2.292840
2017-01-01	2.557756
2016-01-01	1.008417

Listing 9-2 substitutes the missing values with the mean value.

Listing 9-2. Substitute Missing Values with Mean Value

```
df["inflation_cpi"] = df["inflation_cpi"].fillna(df["inflation_cpi"].
mean())
```

Descriptive Statistics

Listing 9-3 calculates Great Britain's inflation and consumer prices from 1960 to 2020. Figure 9-1 shows the plot of this data.

Listing 9-3. Great Britain's Inflation and Consumer Prices Line Plot

```
df["inflation_cpi"].plot(kind="line",color="green",lw=4)
plt.title("Great Britain's inflation, consumer prices (annual %)")
plt.ylabel("Inflation, consumer prices (annual %)")
plt.xlabel("Date")
plt.legend(loc="best")
plt.show()
```

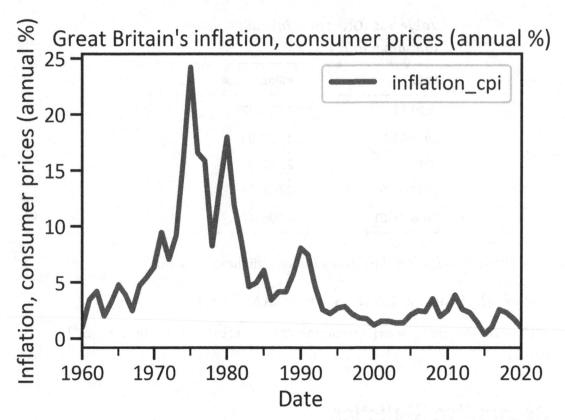

Figure 9-1. *Great Britain's inflation and consumer prices line plot*

Figure 9-1 shows a sharp increase in the early years of 1970 (when Great Britain's inflation reached 24.207288). However, in the mid-1970s, there was a noticeable decrease. In 2015, Great Britain's inflation reached a low of 0.368047. Listing 9-4 calculates the distribution of Great Britain's inflation and consumer prices (see Figure 9-2).

Listing 9-4. Great Britain's Inflation and Consumer Prices Distribution

```
df["inflation_cpi"].plot(kind="hist",color="green")
plt.title("Great Britain's inflation, consumer prices (annual %)")
plt.ylabel("Inflation, consumer prices (annual %)")
plt.xlabel("Date")
plt.legend(loc="best")
plt.show()
```

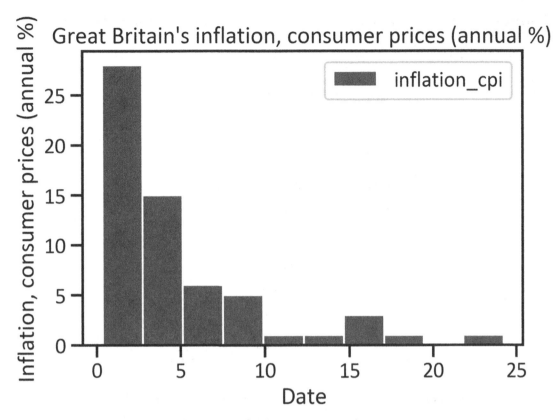

Figure 9-2. *Loading Great Britain's inflation and consumer prices distribution*

Figure 9-2 shows that this data is positively skewed. Note that the model does not require any assumptions regarding the structure of the data (the model assumes the variable is unrecognized). The command in Listing 9-5 generates a table that provides more details about the central tendency and dispersion of the data related to Great Britain's inflation and consumer prices (see Table 9-3).

Listing 9-5. Great Britain's Inflation and Consumer Prices Descriptive Summary

```
df.describe()
```

Table 9-3. *Descriptive Summary*

Variable	inflation_cpi
Count	61.000000
Mean	5.090183
Std	4.858125
Min	0.368047
25%	2.089136
50%	3.427609
75%	6.071394
Max	24.207288

Table 9-3 shows that the mean value of Great Britain's inflation and consumer prices is 5.090183%. It also shows that independent data points deviate from the mean value by 4.858125. The minimum value is 0.368047 and the maximum is 24.207288.

Monte Carlo Simulation Model Development

The benefit of applying simulators is that they are computationally inexpensive—they train large samples faster. Listing 9-6 develops the Monte Carlo simulation model using the `pandas_montecarlo` library.

Listing 9-6. The Monte Carlo Model on Great Britain's Inflation and Consumer Prices

```
import pandas_montecarlo
mc = df['inflation_cpi'].montecarlo(sims=10, bust=-0.1, goal=1)
```

Listing 9-6 specifies the *bust*, which is the probability of going bust as 0.1, and the *goal*, which is the probability of attaining a 100% goal.

Simulation Results

Figure 9-3 shows the predicted output values of Great Britain's inflation and consumer prices across several occurrences, based on the code in Listing 9-7.

Listing 9-7. Great Britain's Inflation and Consumer Prices: Monte Carlo Simulation Results

```
mc.plot(title="Great Britain's inflation, consumer prices (annual %)
simulation results")
```

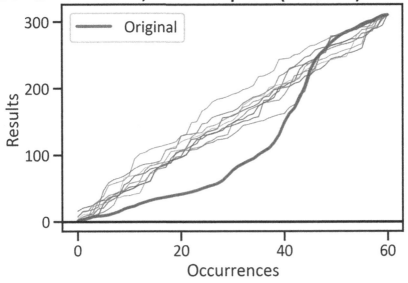

Figure 9-3. *Monte Carlo simulation results on Great Britain's inflation and consumer prices*

Figure 9-3 shows that Great Britain's inflation and consumer prices surge upward across multiple trials.

Simulation Distribution

Table 9-4 outlines the central tendency and dispersion of Great Britain's inflation and consumer prices simulation results based on the code in Listing 9-8.

Listing 9-8. Great Britain's Inflation and Consumer Prices: Monte Carlo Simulation Descriptive Summary

```
simulation_results = pd.DataFrame(mc.data)
simulation_results.describe()
```

Table 9-4. Great Britain's Inflation and Consumer Prices: Monte Carlo Simulation Descriptive Summary

	Original	1	2	3	4	5	6	7	8	9
Count	61.000000	61.000000	61.000000	61.000000	61.000000	61.000000	61.000000	61.000000	61.000000	61.000000
Mean	5.090183	5.090183	5.090183	5.090183	5.090183	5.090183	5.090183	5.090183	5.090183	5.090183
Std	4.858125	4.858125	4.858125	4.858125	4.858125	4.858125	4.858125	4.858125	4.858125	4.858125
Min	0.368047	0.368047	0.368047	0.368047	0.368047	0.368047	0.368047	0.368047	0.368047	0.368047
25%	2.089136	2.089136	2.089136	2.089136	2.089136	2.089136	2.089136	2.089136	2.089136	2.089136
50%	3.427609	3.427609	3.427609	3.427609	3.427609	3.427609	3.427609	3.427609	3.427609	3.427609
75%	6.071394	6.071394	6.071394	6.071394	6.071394	6.071394	6.071394	6.071394	6.071394	6.071394
Max	24.207288	24.207288	24.207288	24.207288	24.207288	24.207288	24.207288	24.207288	24.207288	24.207288

197

For better comprehension of the distribution of Great Britain's inflation and consumer prices simulation results, see Figures 9-4 and 9-5. Listing 9-9 determines the first five simulation results.

Listing 9-9. The First Five Simulation Results Distribution

```
simulation_results.iloc[::,1:6].plot(kind="box",color="green")
plt.title("1st 5 simulation results distribution")
plt.ylabel("Values")
plt.show()
```

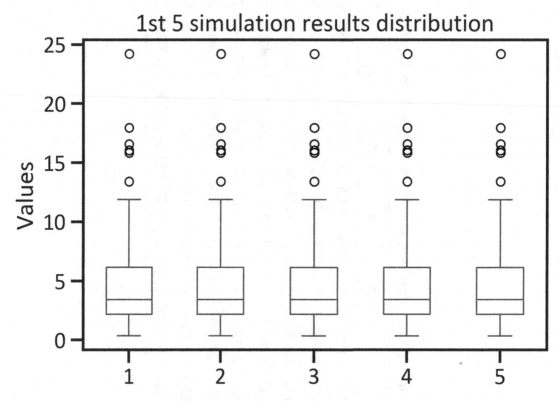

Figure 9-4. *First five simulation results distribution*

Figures 9-4 and 9-5 shows that all distributions of the simulation results are near normal and there are six outliers in each distribution. See Listing 9-10.

Listing 9-10. Other Great Britain's Inflation and Consumer Prices: Simulation Results Distribution

```
simulation_results.iloc[::,6:10].plot(kind="box",color="green")
plt.title("Other simulation results distribution")
plt.ylabel("Values")
plt.show()
```

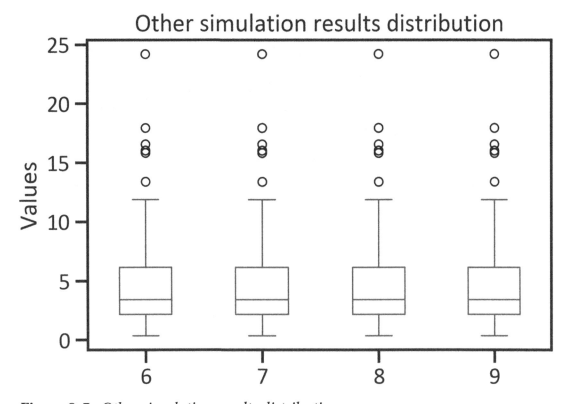

Figure 9-5. *Other simulation results distribution*

Figures 9-4 and 9-5 show that the simulation results saturate the lower echelon with a visible presence of six outliers. Be sure to substitute for the outliers before developing a Monte Carlo simulation model. Listing 9-11 returns Table 9-5, which shows a straightforward way of retrieving Monte Carlo simulation results statistics. These include the *max drawdown,* which is the extent to which Great Britain's inflation and consumer price index drops from its peak before recovering,, the *bust,* which is the probability of going bust, and the goal, which is the probability of attaining a 100% goal.

Listing 9-11. Test Results

```
pd.DataFrame(pd.Series(mc.stats))
```

Table 9-5. *Test Results*

	0
Min	3.105012e+02
Max	3.105012e+02
Mean	3.105012e+02
Median	3.105012e+02
Std	9.087049e-14
Maxdd	NaN
Bust	0.000000e+00
Goal	1.000000e+00

CHAPTER 10

Economic Causal Analysis Applying Structural Equation Modeling

This last chapter of a book introduces ways of approaching economic problems by applying various quantitative techniques. It covers the structural equation model (SEM). Note that this is not a single method, but a framework that includes multiple methods. This method serves many purposes. It applies covariance analysis, for investigating joint variability, correlation analysis, for investigating statistical dependence, factor analysis, for investigating the variability explained by the data about the model, and multivariate regression analysis, for investigating how various predictor variables influence the response variable. It also uses path analysis, for investigating structural models without measurement models.

The structural equation models include a set of models that determine the nature of causal relationships among sets of variables. It includes factor analysis, path analysis, and regression analysis. It helps developers investigate mediating relationships, so they can detect how the presence of other variables weakens or strengthens the nature of the structural relationship between the predictor variable and the response variable.

SEM has an advantage over the models previously covered, as it investigates multiple relationships to predict multiple response variables at the same time. Figure 10-1 shows how the models covered in the preceding chapters work. Most models focus on the direct influence of the predictor variables on a response variable.

© Tshepo Chris Nokeri 2022
T. C. Nokeri, *Econometrics and Data Science*, https://doi.org/10.1007/978-1-4842-7434-7_10

Figure 10-1. *Machine learning model*

Figure 10-1 shows that *x* has a direct influence on *y*. A function may be used to operate on a predictor variable (*x*) and generate an output value (*y*). When there is more than one response variable, the model experiences problems. It is also tedious to estimate multiple equations separately. This is where SEM comes in. SEM handles more than the predictor variable. In addition, it enables developer to consider the mediating effects of other variables.

Framing Structural Relationships

SEM enables you to investigate direct and indirect effects. Figure 10-2 shows that the variable *m* mediates the structural relationship between *x* and *y*. The mediator helps determine whether *m* weakens or strengthens the relationship.

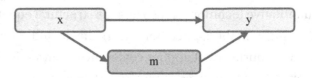

Figure 10-2. *Structural equation model*

In Figure 10-2, *x* represents the predictor variable, *y* represents the response variable, and *m* represents the mediating variable. SEM will study the direct effect of *x* on *y* and the indirect effects of *m* on the relationship between *x* and *y* at the same time. This will enable you to have multiple equations.

Context of This Chapter

Figure 10-3 demonstrates the hypothetical framework that this chapter explores.

The primary research questions are:

- To what extent does Swedish GDP per capita growth (as an annual percentage) influence the final consumption expenditure (in current U.S. dollars)?

- How does Swedish inflation/consumer price index (as a percentage) impact the final consumption expenditure (in current U.S. dollars)?

The secondary research questions include:

- Does Swedish life expectancy at birth (in years) affect the relationship between GDP per capita growth (as an annual percentage) and the final consumption expenditure (in current U.S. dollars)?

Based on these research questions, the research hypotheses are as follows:

H_0: Swedish inflation/consumer price index (as a percentage) do not impact final consumption expenditure (in current U.S. dollars).

H_A: Swedish inflation/consumer price index (as a percentage) impact the final consumption expenditure (in current U.S. dollars).

H_0: Swedish GDP per capita growth (as an annual percentage) does not impact the final consumption expenditure (in current U.S. dollars).

H_A: Swedish GDP per capita growth (as an annual percentage) impacts final consumption expenditure (in current U.S. dollars).

H_0: Swedish life expectancy does not affect the relationship between GDP per capita growth (as an annual percentage) and the final consumption expenditure (in current U.S. dollars).

H_A: Swedish life expectancy does affect the relationship between GDP per capita growth (as an annual percentage) and final consumption expenditure (in current U.S. dollars).

Table 10-1 outlines the chapter's macroeconomic indicators.

Table 10-1. *Swedish Macroeconomic Indicators for This Chapter*

Code	Title
FP.CPI.TOTL.ZG	Swedish inflation/consumer price index (as a percentage)
NY.GDP.PCAP.KD.ZG	Swedish GDP per capita growth (as an annual percentage)
SP.DYN.LE00.IN	Swedish life expectancy (in years)
NE.CON.TOTL.CD	Swedish final consumption expenditure (in current U.S. dollars)

Theoretical Framework

Figure 10-3 provides the framework for this problem.

Before you proceed, be sure that you have the semopy library installed in your environment. To install semopy in a Python environment, use pip install semopy.

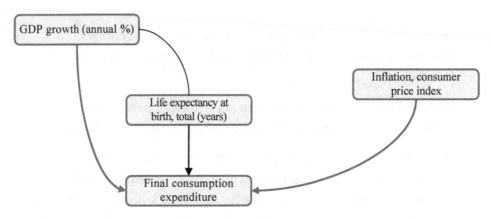

Figure 10-3. *Hypothetical framework*

Final Consumption Expenditure

Figure 10-4 shows the market value of all the goods and services bought by Swedish households. This includes durable products such as cars and personal computers. See Listing 10-1.

Listing 10-1. Swedish Final Consumption Expenditure Line Plot

```
country = ["SWE"]
indicator = {"NE.CON.TOTL.CD":"final_con"}
final_con = wbdata.get_dataframe(indicator, country=country,
convert_date=True)
final_con.plot(kind="line",color="green",lw=4)
plt.title("Swedish final consumption expenditure")
plt.ylabel("Final consumption expenditure")
plt.xlabel("Date")
plt.show()
```

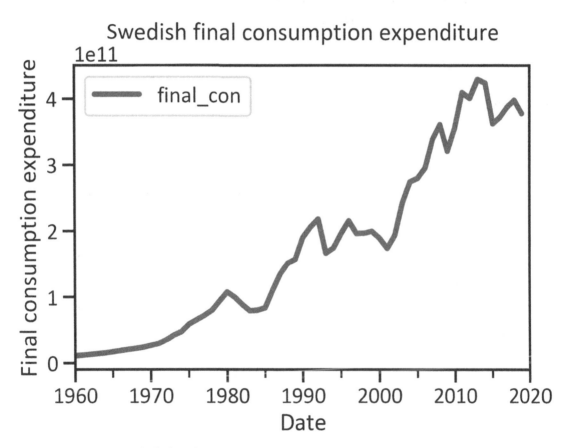

Figure 10-4. *Swedish final consumption expenditure line plot*

Figure 10-4 shows a persistent upward growth in the Swedish final consumption expenditure since 1960. It was at its lowest level in 1960 and reached its peak in 2015.

Inflation and Consumer Prices

Inflation and consumer prices together are used to estimate inflation. This is estimated in part by the annual changes to the consumer price index (an estimate of the cost that average consumer pays for goods or services). It is an exemplary indicator for determining the cost of living in a specific country. Listing 10-2 and Figure 10-5 show annual Swedish inflation and consumer prices.

Listing 10-2. Swedish Inflation and Consumer Prices

```
country = ["SWE"]
indicator = {"FP.CPI.TOTL.ZG":"inflation"}
inflation = wbdata.get_dataframe(indicator, country=country,
convert_date=True)
inflation.plot(kind="line",color="orange",lw=4)
plt.title("Swedish inflation, consumer prices (annual %)")
plt.ylabel("Inflation, consumer prices (annual %)")
plt.xlabel("Date")
plt.show()
```

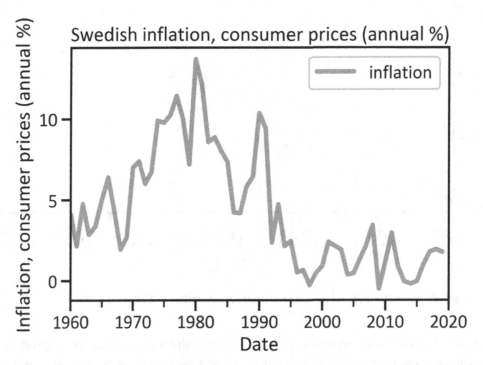

Figure 10-5. *Swedish inflation and consumer prices (as an annual percentage)*

206

Figure 10-5 shows that Swedish inflation and consumer prices tripled from the late 1960s to early 1980. This indicator reached its highest peak in 1982 (at 15%).

Life Expectancy in Sweden

Figure 10-6 shows the average number of years a newborn Swedish baby is predicted to live, assuming that the current mortality pattern remains the same in the future. See Listing 10-3.

Listing 10-3. Swedish Life Expectancy (in Years)

```
country = ["SWE"]
indicator = {"SP.DYN.LE00.IN":"life_exp"}
life_exp = wbdata.get_dataframe(indicator, country=country,
convert_date=True)
life_exp.plot(kind="line",color="navy",lw=4)
plt.title("Swedish life expectancy at birth, total (years)")
plt.ylabel("Life expectancy at birth, total (years)")
plt.xlabel("Date")
plt.show()
```

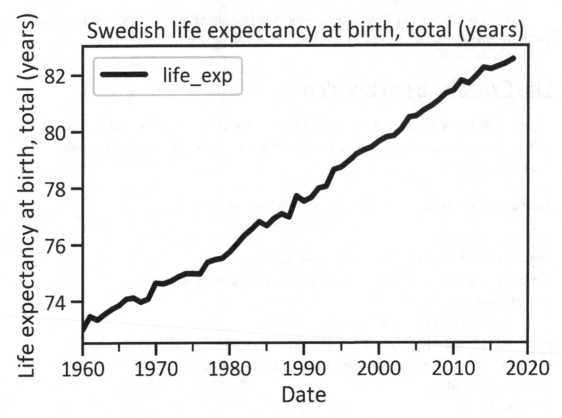

Figure 10-6. *Swedish life expectancy at birth (in years)*

Figure 10-6 shows an extreme upward trend in Swedish life expectancy. In 1960, the average life expectancy was 74 years. As of 2020, that number was beyond the age of 82.

GDP Per Capita Growth

GDP per capita growth is the annual GDP divided by the population of an economy. Figure 10-7 demonstrates Swedish GDP per capita growth (as an annual percentage). See Listing 10-4.

Listing 10-4. Swedish GDP Per Capita Growth (as an Annual Percentage)

```
country = ["SWE"]
indicator = {"NY.GDP.PCAP.KD.ZG":"gdp"}
gdp = wbdata.get_dataframe(indicator, country=country, convert_date=True)
gdp.plot(kind="line",color="red",lw=4)
plt.title("Swedish GDP per capita growth (annual %)")
plt.ylabel("GDP per capita growth (annual %)")
plt.xlabel("Date")
plt.show()
```

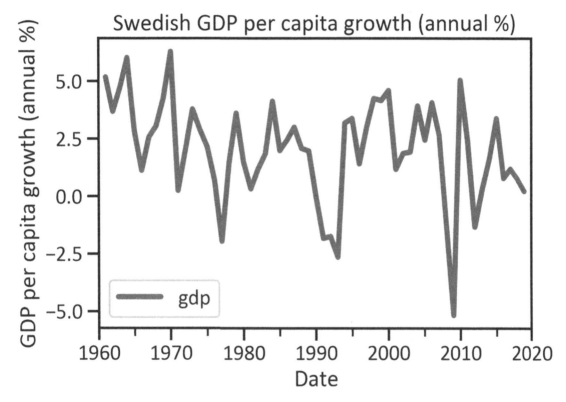

Figure 10-7. *Swedish GDP per capita growth (as an annual percentage)*

Figure 10-7 shows that Swedish GDP per capita growth is unstable. It reached its lowest peak in 2019 (at 5.151263) and its highest peak in 1970 (at 6.292338). Listing 10-5 retrieves all the Swedish macroeconomic data for this chapter and Listing 10-6 shows a descriptive summary (see Table 10-2).

Listing 10-5. Load Swedish Macroeconomic Indicators

```
country = ["SWE"]
indicator = {"NE.CON.PRVT.ZS":"final_con",
             "FP.CPI.TOTL.ZG":"inflation",
             "SP.DYN.LE00.IN":"life_exp",
             "NY.GDP.PCAP.KD.ZG":"gdp"}
df = wbdata.get_dataframe(indicator, country=country, convert_date=True)
```

Listing 10-6. Descriptive Summary

```
df.describe().transpose()
```

Table 10-2. Descriptive Summary

	Count	Mean	Std	Min	25%	50%	75%	Max
final_con	61.0	48.881491	2.647688	44.307210	46.935678	48.399953	50.415939	55.829462
inflation	61.0	4.297861	3.673401	-0.494461	1.360215	2.961151	7.016359	13.706322
life_exp	60.0	77.766451	3.015892	73.005610	74.977683	77.601829	80.509756	82.958537
gdp_per_capita	60.0	1.931304	2.285016	-5.151263	0.790254	2.040267	3.446864	6.292338

Table 10-2 shows that, for Sweden:

- The mean final consumption expenditure is $48.881491.

- The mean inflation/consumer price index is 4.297861%.

- The mean life expectancy is 77.766451 years.

- The mean GDP per capita growth is 1.931304%.

- Independent data points of the final consumption expenditure deviate from the mean value by 2.647688.

- The inflation/consumer price index deviate by 3.673401.

- The life expectancy deviates by 3.015892 years.

- The GDP growth deviates by 2.285016%.

Covariance Analysis

Listing 10-7 investigates joint variability among variables (see Table 10-3).

Listing 10-7. Covariance Matrix

```
dfcov = df.cov()
dfcov
```

Table 10-3. *Covariance Matrix*

	final_con	**inflation**	**life_exp**	**gdp_per_capita**
final_con	7.010251	3.467993	-7.010302	1.559997
inflation	3.467993	13.493871	-6.615968	-0.905779
life_exp	-7.010302	-6.615968	9.095604	-1.771163
gdp_per_capita	1.559997	-0.905779	-1.771163	5.221296

Table 10-3 outlines the estimated covariance of the set of variables you retrieved.

Correlation Analysis

Figure 10-8 shows the statistical dependence among the variables, which is determined from Listing 10-8.

Listing 10-8. Pearson Correlation Matrix

```
import seaborn as sns
dfcorr = df.corr(method="pearson")
sns.heatmap(dfcorr, annot=True, annot_kws={"size":12},cmap="Blues")
plt.title("Pearson correlation matrix")
plt.show()
```

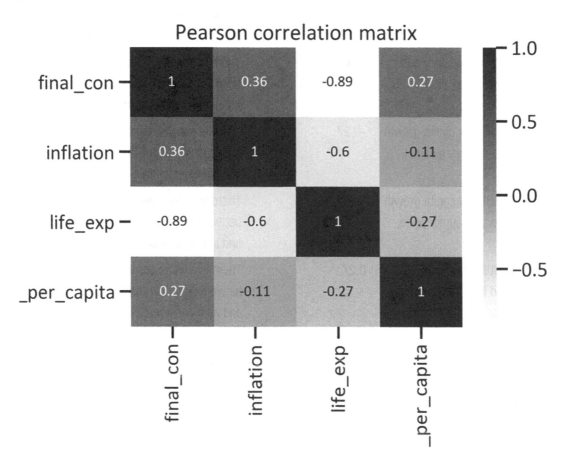

Figure 10-8. *Pearson correlation matrix*

Table 10-4 interprets the results in Figure 10-8.

Table 10-4. *Interpretation of Pearson Correlation Coefficients*

Variables	Pearson Correlation Coefficient	Findings
Swedish inflation/consumer price index and final consumption expenditure	0.36	There is a weak positive correlation between Swedish inflation/consumer price index and final consumption expenditure.
Swedish lending interest rate and final consumption expenditure	-0.89	There is an extreme negative correlation between Swedish lending interest rate and final consumption expenditure.
Swedish life expectancy and inflation/consumer price index	-0.6	There is a moderate negative correlation between Swedish life expectancy and inflation/consumer price index.
Swedish GDP per capita growth and life expectancy (years)	-0.27	There is a weak negative correlation between Swedish GDP per capita growth and life expectancy.
Swedish GDP per capita growth and inflation/consumer price index	-0.11	There is a weak negative correlation between Swedish GDP per capita growth and inflation/consumer price index.
Swedish GDP per capita growth and final consumption expenditure	0.27	There is a weak positive correlation between Swedish GDP per capita growth and final consumption expenditure.

After discovering the correlation between these variables, you can create a plot to graphically represent the correlation relationships (using the command in Listing 10-9). Figure 10-9 shows the resulting pair plots.

Listing 10-9. Pair plot

```
sns.pairplot(df)
```

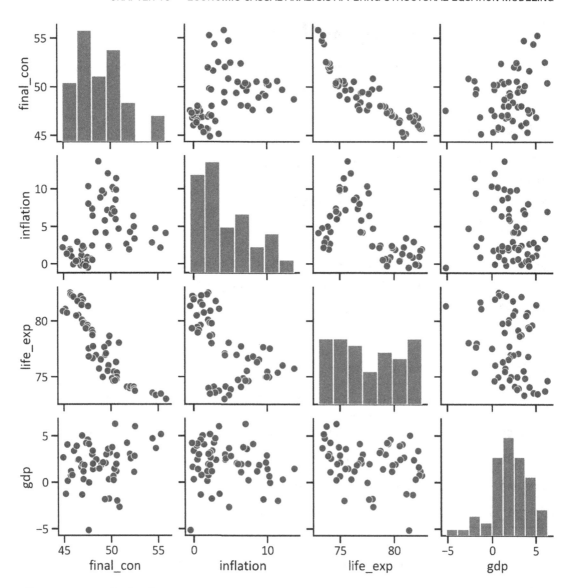

Figure 10-9. *Pair plots*

Figure 10-9 shows histograms of all the variables and scatter plots of the relationships among variables in the data.

Correlation Severity Analysis

Listing 10-10 determines the severity of each dependency by applying the Eigen matrix (see Table 10-5). Eigenvalues are a measure of maximum variances.

Listing 10-10. Eigen Matrix

```
eigenvalues, eigenvectors = np.linalg.eig(dfcorr)
eigenvalues = pd.DataFrame(eigenvalues)
eigenvalues.columns = ["Eigen values"]
eigenvectors = pd.DataFrame(eigenvectors)
eigenvectors.columns = df.columns
eigenvectors = pd.DataFrame(eigenvectors)
eigenmatrix = pd.concat([eigenvalues, eigenvectors],axis=1)
eigenmatrix.index = df.columns
eigenmatrix
```

Table 10-5. *Eigen Matrix*

	Eigen Values	final_con	inflation	life_exp	gdp_per_capita
final_con	2.323484	0.594607	0.595973	0.527068	0.116012
inflation	0.059704	0.432198	0.262541	-0.662923	-0.552096
life_exp	0.512778	-0.641591	0.755077	-0.133840	0.017516
gdp_per_capita	1.104034	0.219109	0.075813	-0.514606	0.825484

Based on Table 3-4, there is no multicollinearity in the data. The eigenvalues are all below 3.

Structural Equation Model Estimation

SEM operates on variables by applying the Maximum Likelihood method or the general least-squares method. The Maximum Likelihood method works when the implicit assumptions of regression are not satisfied and it determines the parameters that best fit the data. The general least-squares method regresses the variables. Listing 10-11 structures the model's hypothetical framework.

Listing 10-11. Develop Hypothetical Structure

```
import semopy
import wbdata
from semopy import Model
import semopy
import wbdata
from semopy import Model
mod = """   life_exp ~ gdp_per_capita
            final_con ~ inflation + life_exp + gdp_per_capita
      """
```

Structural Equation Model Development

Listing 10-12 trains SEM with the default hyperparameters.

Listing 10-12. Structural Equation Model

```
m = Model(mod)
m.fit(df)
SolverResult(fun=0.549063553713184, success=True, n_it=18,
x=array([-0.33778359, -0.19991271, -0.92551867, -0.04842324,  8.35819007,
       1.2715286 ]), message='Optimization terminated successfully',
       name_method='SLSQP', name_obj='MLW')
```

These results show that SEM applies the Maximum Likelihood method, including the objective function.

Structural Equation Model Information

Listing 10-13 retrieves the information about SEM relating to the objective name, optimization method, objective value, and the number of iterations.

Listing 10-13. Model Information

```
print(m.fit(df))
Name of objective: MLW
Optimization method: SLSQP
Optimization successful.
Optimization terminated successfully
Objective value: 0.549
Number of iterations: 1
Params: -0.338 -0.200 -0.926 -0.048 8.358 1.272
```

Listing 10-14 returns the values that SEM predicts and then tabulates those values (see Table 10-6).

Listing 10-14. Make Predictions

```
preds = m.predict(df)
preds
```

Table 10-6. *Swedish Actual and Predicted Final Consumption Expenditure*

Date	final_con	gdp_per_capita	Inflation	life_exp
2020-01-01	44.307210	-3.517627	0.497367	-40.408627
2019-01-01	45.287253	0.347439	1.784151	82.958537
2018-01-01	45.697290	0.772577	1.953535	82.558537
2017-01-01	45.718184	1.195148	1.794499	82.409756
2016-01-01	45.856915	0.796146	0.984269	82.307317
...	...	...	...	...
1964-01-01	52.543020	6.026059	3.387662	73.733171
1963-01-01	54.398699	4.735940	2.871740	73.555366
1962-01-01	54.746401	3.685605	4.766197	73.350488
1961-01-01	55.275500	5.184619	2.157973	73.474390
1960-01-01	55.829462	-36.394777	4.141779	73.005610

Figure 10-10 shows the actual and predicted values of Swedish final consumption expenditure. See Listing 10-15.

Listing 10-15. Swedish Actual and Predicted Final Consumption Expenditures

```
preds["final_con"].plot(lw=4)
df["final_con"].plot(lw=4)
plt.title("Swedish actual and predicted FCE")
plt.xlabel("Date")
plt.ylabel("FCE")
plt.show()
```

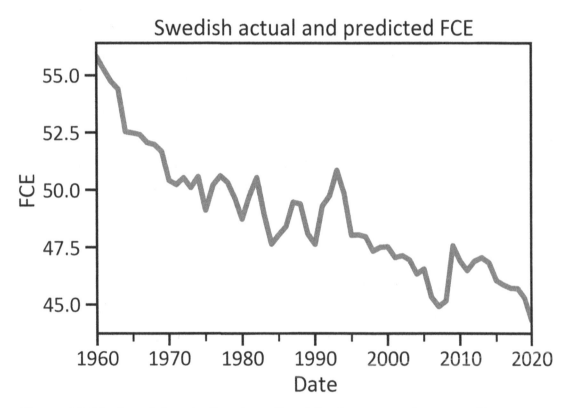

Figure 10-10. *Swedish actual and predicted final consumption expenditures*

Figure 10-10 demonstrates that SEM is skillful in predicting the Swedish final consumption expenditure.

Structural Equation Model Inspection

Listing 10-16 retrieves the estimates, the standard error, the z-value, and the p-value, then tabulates this information in Table 10-7. You can apply the p-value to determine the significance of the hypothetical claims made earlier in the chapter. In addition, use the standard error to determine the magnitude of errors.

Listing 10-16. Structural Equation Model Inspection

```
m.inspect()
```

Table 10-7. *Structural Equation Model Profile Table*

	lval	op	rval	Estimate	Std. Err	z-value	p-value
0	life_exp	~	gdp_per_capita	-0.337784	0.163362	-2.067697	3.866852e-02
1	final_con	~	inflation	-0.199913	0.039862	-5.015087	5.300940e-07
2	final_con	~	life_exp	-0.925519	0.049939	-18.532890	0.000000e+00
3	final_con	~	gdp_per_capita	-0.048423	0.066274	-0.730648	4.649940e-01
4	life_exp	~~	life_exp	8.358190	1.513430	5.522681	3.338665e-08
5	final_con	~~	final_con	1.271529	0.230238	5.522681	3.338665e-08

Table 10-7 shows that all the relationships are significant (refer to the "Visualize Structural Relationships" section to better interpret these results). To further study how well SEM performs, use one of the following:

- Absolute Fit Index (AFI) determines whether the model fits the data by applying tests like the Likelihood Ratio test. When applying these indices, focus mostly on the χ^2 (Chi-Square) value.

- Incremental Fit Index includes the Normed-Fit Index (NFI), which compares the $\chi2$ value of the model to the $\chi2$ of the null model, and the Comparative Fit Index (CFI), which compares the $\chi2$ value of the model to the $\chi2$ of the null model, considering the sample size.

Report Indices

Listing 10-17 retrieves the report indices (see Table 10-8).

Listing 10-17. Report Indices

```
stats = semopy.calc_stats(m)
stats.transpose()
```

Table 10-8. Report Indices

	DoF	DoF Baseline	chi2	chi2 p-value	chi2 Baseline	CFI	GFI	AGFI	NFI	TLI	RMSEA	AIC	BIC	LogLik
Value	4	8	36.813532	1.9678669e-07	172.658646	0.800718	0.786784	0.573569	0.786784	0.601436	0.369761	10.792999	23.458242	0.603501

Table 10-8 shows that SEM loses about 11% of the information when it explains the data generation process (see the AIC, which tells us much about the quality of model). The degrees of freedom is 4 and $\chi 2$ is 36.813532. The most important index is the $\chi 2$ p-value. We apply it to decide whether to accept or reject the following hypotheses:

H_0: The results are not statistically significant.

H_A: The results are statistically significant.

Table 10-8 shows indicates that the $\chi 2$ p-value is 0.0000001967869, which is less than the 0.05. That means we reject the null hypothesis in favor of the alternative. The results *are* statistically significant.

Visualize Structural Relationships

Figure 10-11 shows the relationship among variables. It also plots the covariance estimates and the p-value. Use the command in Listing 10-18 to plot this information.

Listing 10-18. Visualize Structural Relationships

```
g = semopy.semplot(m, "semplot.png", plot_covs=True)
g
```

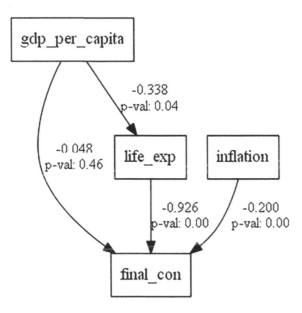

Figure 10-11. *Structural equation model plot*

Figure 10-11 shows that:

- Swedish inflation/consumer price index impacts the final consumption expenditure. They vary by -0.200.

- Swedish GDP per capita growth impacts the final consumption expenditure. They vary by -0.048.

- Swedish life expectancy strengthens the relationship between GDP per capita growth and the final consumption expenditure.

- Inflation/consumer price index varies with life expectancy by -0.338 and life expectancy varies with final consumption expenditure by -0.926.

Conclusion

This chapter introduced SEM. It explored the structural relationships between Swedish GDP per capita growth, inflation/consumer price index, and final consumption expenditure. In addition, it explored the mediating effect of Swedish life expectancy on the relationship between GDP per capita growth and final consumption expenditure.

First, it explained how to develop a theoretical framework that you can test by applying the method. Second, it covered how to determine joint variability by applying covariance, how to determine association by applying the Pearson correlation, and how to determine the severity of correlation by applying Eigen matrix. You saw how to reduce the data by applying the principal component analysis. Lastly, you developed SEM and found that the results are significant.

Index

A

Absolute Fit Index (AFI), 220
Actual data points, 32, 34–37, 75, 76
Additive model, 93
    development, 94
    forecast, 95
Area under the curve (AUC), 125
Artificial neural network, 179
    data reprocessing, 179
    structuring, 179, 180
    wrapping, 180
Atlas method, 98, 101, 179
Augmented Dickey-Fuller test, 83, 88, 89
Autocorrelation function, 39, 40, 78, 90, 91
Autoregressive Integrated Moving
        Average, 93
Autoregressive Moving Average (ARIMA), 83

B

Boltzmann machine classifier, 4, 5,
        164–169, 179, 181, 188
Box plots, 20

C

Classification report, 123, 166, 167, 174
Comparative Fit Index (CFI), 221
Conda environment, 16, 21, 83, 84
Confusion matrix, 121–123, 165

Confusion matrix interpretation, 123, 165,
    166, 173
Consumer prices, 190, 198, 206
Conventional quantitative models, 4
Convolutional Neural Network, 5
Correlation analysis, 13, 25, 61, 90, 113,
    201, 213
Covariance analysis, 13, 24, 25
Covariance matrix, 25, 60, 112, 212
Cross-validation loss, 185–187
cross_val_score() method, 31
Cut-off point, 153

D

Data generation process, 223
Data manipulation, 163
Data preprocessing, 68, 120
Data sources, 2, 6, 8, 9
Deep learning, 1, 4, 5, 128
Dimension reduction technique, 45,
    65, 66
Downward trend, 83, 129

E

Econometrics, 1–3, 128
Economic design, 2
Eigen loadings, 63
Eigen matrix, 45, 64, 116, 216
Eigen method, 63

Printed in the United States
by Baker & Taylor Publisher Services